Do
Brilliantly

AS Maths

John Berry, Ted Graham & Roger Williamson

Series Editor: Jayne de Courcy

Published by HarperCollins*Publishers* Limited
77–85 Fulham Palace Road
London W6 8JB

www.**Collins**Education.com
On-line support for schools and colleges

© HarperCollins*Publishers* Ltd 2001

First published 2001

Reprinted 2001

ISBN 0 00710702 1

John Berry, Ted Graham and Roger Williamson assert the moral right to be identified as the authors of this work.

British Library Cataloguing in Publication Data
A catalogue record for this book is available from the British Library

Edited by Joan Miller
Production by Kathryn Botterill
Cover design by Susi Martin-Taylor
Book design by Gecko Limited
Printed and bound in Scotprint.

Acknowledgements
The Authors and Publishers are grateful to the following for permission to reproduce copyright material:
The following questions: p.14 Q3, p.14 Q10, p.18 Q1, p.18 Q6, p.18 Q7, p.18 Q10, p.26 Q7, p.26 Q8, p.30 Q5, p.30 Q8, p.35 Q10, p.48 Q6, p.54 Q2, p.60 Q5, p.67 Q2, p.71 Q3, p.72 Q5, p.84 Q2, p.84 Q3, p.88 Q3, p.95 Q2, p.95 Q4, p.95 Q6 reproduced with the permission of the Assessment and Qualifications Alliance **AQA**
The following questions: p.77 Q1, p.77 Q2, p.77 Q3, p.79 Q5, p.80 Q5, p.83 Q3, p.85 Q5, p.89 Q1, p.95 Q3, p.95 Q5, p.101 Q3, p.102 Q5 reproduced with the permission of the Assessment and Qualifications Alliance **AQA (AEB)**
Question 4 on p.56 reproduced with the permission of the Assessment and Qualifications Alliance **AQA (SMP)**
The following questions: p.10 Q5, p.14 Q5, p.18 Q9, p.21 Q5, p.26 Q10, p.31 Q9, p.37 Q1, p.38 Q3, p.52 Q8, p.52 Q9, p.56 Q2, p.60 Q2, p.60 Q2, p.60 Q3, p.64 Q6, p.66 Q1, p.69 Q4, p.94 Q3, p.95 Q1, p.102 Q4 reproduced with the permission of **Edexcel**
The following questions: p.10 Q2, p.10 Q3, p.10 Q6, p.18 Q4, p.18 Q8, p.26 Q9, p.30 Q1, p.35 Q9, p.37 Q2, p.39 Q4, p.52 Q7, p.55 Q3, p.56 Q3, p.68 Q3, p.70 Q1, p.75 Q4, p.89 Q4 reproduced with the permission of Oxford Cambridge RSA Examinations **OCR/UCLES**
The following questions: p.40 Q6, p.82 Q2, p.85 Q6, reproduced with the permission of the Welsh Joint Education Committee/Cyd-bwyllgor Addysg Cymru **WJEC**

Note: The Awarding Bodies listed above accept no responsibility whatsoever for the accuracy or method of working in the answers given, which are solely the responsibility of the author and publishers.

Illustrations
Cartoon Artwork – Roger Penwill
DTP Artwork – Jeff Edwards

Every effort has been made to contact the holders of copyright material, but if any have been inadvertently overlooked, the Publishers will be pleased to make the necessary arrangements at

Contents

Pure

How this book will help you........ 4

Algebra and equations I 7

Arithmetic and geometric progressions 11

Trigonometry............................ 15

Coordinate geometry 19

Functions.................................. 22

Differentiation I 27

Integration I.............................. 32

Algebra and equations II 36

Exponentials and logarithms.... 41

Differentiation II 45

Integration II 49

Transformations 53

Mechanics

13 Kinematics on a straight line 57

14 Kinematics and vectors 61

15 Newton's laws and connected particles 65

16 Conservation of momentum 73

Statistics

17 Sampling 76

18 Probability 81

19 Binomial distribution 86

20 Normal distribution 91

21 Correlation and regression........ 96

22 Answers to Questions to try.... 103

How this book will help you

by John Berry, Ted Graham **and** Roger Williamson

Exam practice – how to answer questions better

This book will help you to improve your performance in your AS Maths exam.

The key to success in Maths is lots of practice. In this book we have provided numerous **AS exam questions for you to try to answer, along with support to help you** if you get stuck. The two most common reasons that students lose marks in exams are that they make silly algebraic errors or that they are not familiar with the topics on which the questions are based. We have provided **good coverage of the main AS Maths topics** (Pure, Mechanics and Statistics) so that you can make sure you are familiar with the key topics. We have included two chapters on algebra to help you practise and revise your algebraic skills.

Each chapter in this book is broken down into four separate elements, aimed at giving you as much practice and guidance as possible:

❶ 'Key points to remember' and 'Don't make these mistakes'

On these pages we outline the key points that you need to know for each of the topics. These points should not be new to you and our aim is to provide **a reminder that will help refresh your memory**. We also include **lists of formulae that you must not forget**. In the new AS examination, you are expected to learn formulae rather than be able to look them up, so this section is very important. We also include **checklists of common mistakes that you should try to avoid.**

❷ Exam Questions and Answers with 'How to score full marks'

We have used a number of exam-style questions to illustrate **the methods that you will need to use in your exam**. We provide a question which is followed by a sample correct solution. Alongside this **we provide notes to guide you through the solution and indicate the steps needed to score full marks**.

❸ Questions to try

This is where **you get the opportunity to practise answering the types of question that you will meet in your exam**. Each section starts with some easier questions to warm you up and then includes actual exam questions or exam-style questions. **You need to attempt all the questions as this sort of practice will help you to reach your full potential in your exam.**

❹ Answers to the Questions to try

At the back of the book you will find **solutions to all of the Questions to try**. Try answering the questions as though you were in the exam. Look at the answers if you are really stuck or when you think that you have completed the question correctly. **Alongside each solution we have written a commentary that identifies the key stages in the solution – this will help you if you get stuck**. The commentary will also help you to make sure that you have not omitted any important working that you should have shown.

This book includes questions on Pure Maths, Mechanics and Statistics. **It focuses on the most important or difficult parts of the AS Mathematics core**, which all exam boards must cover for an AS award. Some of the exam boards have chosen to go beyond this core and you will find material for these topics in our A2 Mathematics book. The table below shows how the topics in this book fit into the modules for the different exam boards. Some sections may contain a little more material than is required for your particular exam.

The topics covered by your specification

Chapters in this book	AQA – A	AQA – B	EDEXCEL	OCR – A	OCR – B
1	Me	P1	P1	P1	P1/P2
2	P1	P1/P2	P1	P2	P2
3	P1	P1/P2	P1	P1/P2	P1
4	Me	P1	P1	P1	P1
5	P1	P1	P2	P2	P2
6	Me	P1	P1	P1	P1
7	Me	P1	P1	P1	P1
8	P1	P4	P2	P2	P2
9	P1	P2	P2	P2	P2
10	P1	P2	P2	P2	P2
11	P1	P2	P2	P2	P2
12	P1	P1	P2	P1	P2
13	M1	M1	M1	M1	M1
14	M1	M1	M1		M1
15	M1	M1	M1	M1	M1
16	M1	M1	M1	M1	
17	S1	S1/S2	S3	S2	S1
18	Me	S1/S3	S1	S1	S1
19	S1	S1	S2	S1	S1
20	S1	S1	S1	S2	S2
21	S4	S1	S1	S1	S2

- There are two types of marks available to you: **method marks and accuracy marks**. Accuracy marks are awarded for correct working and answers. Method marks are awarded for working that is of the type required, but where there is a reasonably minor error.

  For example, if when solving a quadratic equation you fail to include a negative sign with one of the numbers, you would lose an accuracy mark, but may be awarded a method mark because you have shown that you know how to solve a quadratic equation. **In order to gain method marks, it is very important that you show your working clearly** so that the examiner can give you credit if the answer is wrong. Short comments to explain to the examiner what you are trying to do may help.

- **In 'show that...' questions, make sure that you really do show that the result is true**. You should include all the necessary steps in your working. Students often miss out obvious – but important – steps when answering these types of question.

- **Make sure that you have learned the formulae that you need to know for each topic**. You may like to make a list that you can spend time studying. Imagine sitting in the exam looking at a GP, but being unable to remember the formula that you need to find its sum. Also be aware of which formulae are in the formula book for your exam board and be able to find them when you need them.

- **Be aware of which papers for your board have calculator restrictions**. This will apply to some of the Pure Maths papers. You should know which calculators to take to which exam. A graphics calculator can be an advantage in an exam, but you do need to have spent time getting to know how to use it.

- When you are asked for an exact answer or are asked to 'show that $x = \dfrac{\sqrt{3}}{7}$,'

  or similar, **do not use decimal approximations as you are working**. Always work with surds or fractions until you obtain the required result.

- **Look at the number of marks that are awarded for each part of a question**. This will give you a guide to how much work you have to do. A question that has one mark for an explanation or comment will expect one sentence. Do not write half a page of explanation, as you will be wasting time.

- **Use the time that you have available in the exam productively**. If you know that you are not progressing with a question, stop and move on to another one. Also, always try the later parts of questions, even if you cannot do the first part. Exam questions often use the 'show that...' format so that students can attempt later parts even if they can't do the 'show that...' part.

1 Algebra and equations I

Key points to remember

Techniques

- A surd is a number $a + \sqrt{b}$ where a and b are integers.
- Use the difference of two squares to remove a surd in the denominator of an expression, e.g. for $a + \sqrt{b}$, multiply both denominator and numerator by $a - \sqrt{b}$ since $(a + \sqrt{b})(a - \sqrt{b}) = a^2 - b$.

Quadratic equations

- Any quadratic equation function $ax^2 + bx + c$ can be written in the form $a(x + p)^2 + q$. This is called **completing the square**.
- A quadratic equation of the form $ax^2 + bx + c = 0$ can be solved using the **formula**:

$$x = \frac{-b \pm \sqrt{b^2 - 4ac}}{2a}$$

- The equation $ax^2 + bx + c = 0$ has real roots if $b^2 - 4ac \geqslant 0$.

Formulae you must know

- $x^a x^b = x^{a+b}$
- $x^a \div x^b = x^{a-b}$
- $(x^a)^b = x^{a \times b} = x^{ab}$
- $x^0 = 1$
- $x^{-a} = \dfrac{1}{x^a}$
- Solutions of $ax^2 + bx + c = 0$ are:

$$x = \frac{-b \pm \sqrt{b^2 - 4ac}}{2a}$$

- $(a + \sqrt{b})(a - \sqrt{b}) = a^2 - b$
- $(a + b)(a - b) = a^2 - b^2$

Graphs of quadratic functions

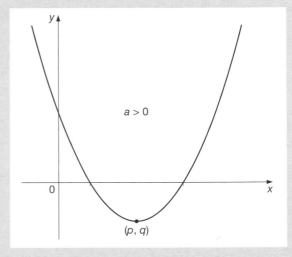

$y = a(x - p)^2 + q$ has a minimum at (p, q).

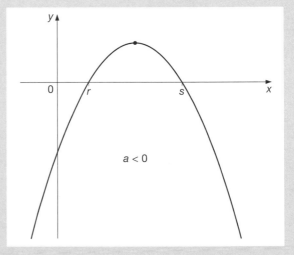

$y = a(x - r)(x - s)$ has roots at $x = r$ and $x = s$.

x^{a+b} is **not** equal to $x^a + x^b$

$(a + b)^2$ is **not** equal to $a^2 + b^2$

x^0 is not equal to 0.

Exam Questions and Student's Answers

How to score full marks

Q1 Express $\dfrac{(\sqrt{3} - 1)^2}{\sqrt{3} + 1}$ in the form $a + b\sqrt{3}$, where a and b are integers.

$$\frac{(\sqrt{3} - 1)^3}{(\sqrt{3} + 1)(\sqrt{3} - 1)} = \frac{(\sqrt{3} - 1)^3}{3 - 1}$$

$$= \frac{(\sqrt{3})^3 - 3(\sqrt{3})^2 + 3\sqrt{3} - 1}{2}$$

$$= \frac{6\sqrt{3} - 10}{2}$$

$$= -5 + 3\sqrt{3}$$

- First, you need to remove the surd in the denominator. Remember that: $(c + \sqrt{d})(c - \sqrt{d}) = c^2 - d$

- You need to multiply numerator and denominator by $\sqrt{3} - 1$.

- Now expand $(\sqrt{3} - 1)^3$ either as a binomial or by expanding $(\sqrt{3} - 1)^2$ first and then multiplying by $(\sqrt{3} - 1)$.

Q2 Find the range of values for x for which $(x - 4)(2x + 3) < 0$.

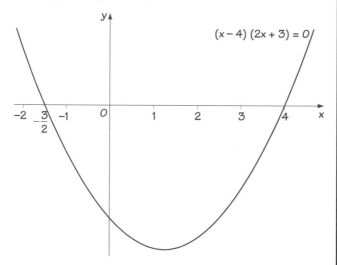

- In these problems, you will usually find that a sketch of the graph showing the x-intercepts helps.

- Remember to look for $<$ or $\leqslant$, or $>$ or $\geqslant$.

$y < 0$ when $-\dfrac{3}{2} < x < 4$

Q3 A curve has equation $y = x^2 + kx + 9$.
  (a) Given that the curve does not cross or touch the x-axis, find the set of values that k can take.
  (b) In the case when $k = 8$, express the equation of the curve in the form $y = (x + a)^2 + b$.

Q3 (c) In the case when $k = -7$, prove that the curve crosses the x-axis at a point $(\alpha, 0)$ where $1.6 < \alpha < 1.7$.

(a) There are no solutions of a quadratic equation when the discriminant $b^2 - 4ac < 0$.

$k^2 - 36 < 0$

- Remember that a positive number has two square roots, so the solution is not simply $k < 6$.

- A sketch graph of $y = k^2 - 36$ helps.

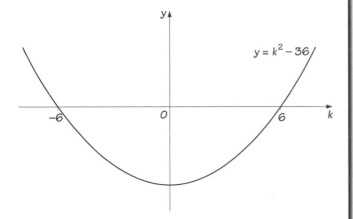

$-6 < k < 6$

(b) $k = 8 \Rightarrow y \Rightarrow x^2 + 8x + 9$

$\Rightarrow y = (x + 4)^2 - 16 + 9$

$\Rightarrow y = (x + 4)^2 - 7$

This is in the form $(x + a)^2 + b$, where $a = 4$, $b = -7$.

- Substitute $k = 8$ into the equation and complete the square.

(c) $k = -7 \Rightarrow y = x^2 - 7x + 9$

$y(1.6) = 1.6^2 - 7 \times 1.6 + 9 = 0.36$

$y(1.7) = 1.7^2 - 7 \times 1.7 + 9 = -0.01$

Since $y(1.6) > 0$ and $y(1.7) < 0$ there is a value α between 1.6 and 1.7 such that $y(\alpha) = 0$.

- Substitute $k = -7$ into the equation, then try solutions.

- Finally, you need to give a clear explanation for your answer.

Q4 (i) Write down the exact value of 7^{-2}.

(ii) Simplify $\dfrac{(x\sqrt{x})^3}{2x^4}$.

(i) $7^{-2} = \dfrac{1}{7^2} = \dfrac{1}{49}$

- You need an exact answer, so don't use your calculator.

(ii) $\dfrac{(x\sqrt{x})^3}{2x^4} = \dfrac{(x^{\frac{3}{2}})^3}{2x^4} = \dfrac{x^{\frac{9}{2}}}{2x^4} = \dfrac{x^{\frac{1}{2}}}{2}$ or $\dfrac{1}{2}\sqrt{x}$

- Simplify or expand the numerator first.

Q1 Simplify each of the following expressions.

(a) $a^5 \times a^{-2}$

(b) $\dfrac{a^5}{a^{-3}}$

(c) $\left(\dfrac{\sqrt[4]{a}}{\sqrt[7]{a}} \right)^3$

Q2 A quadratic function is defined by:

$f(x) = x^2 + kx + 9$

where k is a constant. It is given that the equation $f(x) = 0$ has two distinct real roots. Find the set of values that k can take.

For the case where $k = -4\sqrt{3}$:

(i) express $f(x)$ in the form $(x + a)^2 + b$, stating the values of a and b, and hence write down the least value taken by $f(x)$

(ii) solve the equation $f(x) = 0$, expressing your answer in terms of surds, simplified as far as possible.

Q3 Solve the inequality $x^2 > x + 20$.

Q4 Two consecutive integers are chosen so that their sum is greater than 10 and their product is less than 72. Find the range of values within which the two numbers must be.

Q5 The specification for a new rectangular car park states that its length, x m, is to be 5 m more than its width. The perimeter of the car park is to be greater than 32 m.

(a) Form a linear inequality in x.

(b) The area of the car park is to be less than 104 m^2. Form a quadratic inequality in x.

(c) By solving your inequalities, determine the possible range of values of the length of the car park.

Q6 (i) Solve the simultaneous equations.

$y = x^2 - 3x + 2 \qquad y = 3x - 7$

(ii) Interpret your solution to part (i) geometrically.

Answers can be found on page 103.

Key points to remember

- A **sequence is a list of terms** defined by some relationship,
  e.g. 3, 7, 11, 15, 19, ...

- A **series is the sum of the terms of a sequence**,
  e.g. 3 + 7 + 11 + 15 + 19 + ...

- A **sequence may converge to a limit**,
  e.g. $1, \frac{1}{2}, \frac{1}{4}, \frac{1}{8}, ...$

  which **converges to 0**

  or it may **diverge**,
  e.g. 1, 4, 8, 32, ...
  which **increases to infinity**.

- A **series may converge**,
  e.g. $1 + \frac{1}{3} + \frac{1}{9} + \frac{1}{27} + ...$

  which **converges to** $\frac{3}{2}$

  or it may **diverge**,
  e.g. 1 + 3 + 5 + 7 + 9 + ...
  which **increases to infinity**.

- **Arithmetic progression**

  If the **first term** is a and the **common difference** is d, then the **nth term** u_n is given by: $u_n = a + (n-1)d$ and the **sum of the first n terms** S_n is given by:
  $$S_n = \frac{1}{2}n(2a + (n-1)d) \text{ or } S_n = \frac{1}{2}n(a + l)$$

  where l is the last term of the sequence. The relationship between successive terms is given by:
  $$u_{n+1} = u_n + d$$

- **Geometric progression**

  If the **first term** is a and the **common ratio** is r, then the **nth term** u_n is given by:
  $$u_n = ar^{n-1}$$
  and the **sum of the first n terms** S_n is given by:
  $$S_n = a\left(\frac{r^n - 1}{r - 1}\right) \text{ or } S_n = a\left(\frac{1 - r^n}{1 - r}\right)$$

  The relationship between successive terms is given by:
  $$u_{n+1} = u_n \times r$$

- **Sum to infinity of a GP**

  If the **common ratio r** of a GP is such that $-1 < r < 1$, then the sum to infinity of the GP is given by:
  $$S_\infty = \frac{a}{1 - r}$$

Formulae you must know

- $u_n = a + (n-1)d$
- $S_n = \frac{1}{2}n(2a + (n-1)d)$
- $u_n = ar^{n-1}$
- $S_n = a\left(\frac{r^n - 1}{r - 1}\right) \text{ or } S_n = a\left(\frac{1 - r^n}{1 - r}\right)$
- $S_\infty = \frac{a}{1 - r}$

Don't make these mistakes...

Don't use the **wrong number of terms**.

Don't forget to **subtract 1 from n when finding the nth term**.

Don't use the **sum to infinity formula** without checking that $-1 < r < 1$.

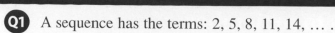
Q1 A sequence has the terms: 2, 5, 8, 11, 14, ...
  (a) Find the 100th term of the sequence.
  (b) Find the sum of the first 100 terms of the sequence.

(a) $u_{100} = 2 + (100 - 1) \times 3$

$\qquad = 2 + 99 \times 3$

$\qquad = 299$

(b) $S_n = \frac{1}{2} \times 100 \times (2 \times 2 + (100 - 1) \times 3)$

$\qquad = 50 \times 301$

$\qquad = 15\,050$

- You need to **recognise that this is an arithmetic progression** because there is a constant difference between adjacent terms. The sequence has common difference 3 and first term 2, so $d = 3$ and $a = 2$.

- **Use the formula** $u_n = a + (n - 1)d$, with $n = 100$.

- **Use the formula** $S_n = \frac{1}{2}n(2a + (n - 1)d)$, with $n = 100$.

Q2 A geometric progression, with positive terms, has first term 4 and third term 2.56.
  (a) Find the sum of the first five terms.
  (b) Find the sum to infinity of the terms of the progression.

(a) $4r^2 = 2.56$

$r = \sqrt{\dfrac{2.56}{4}} = 0.8$

$S_5 = 4\left(\dfrac{1 - 0.8^5}{1 - 0.8}\right) = 13.4$

correct to three significant figures.

(b) $S_\infty = 4 \div (1 - 0.8) = 20$

- Your first task is to **find the common ratio** for the GP. To do this, substitute $a = 4$, $n = 3$ and $u_3 = 2.56$ into the formula $u_n = ar^{n-1}$ and solve for r.

- Then **use the formula** $S_n = a\left(\dfrac{1 - r^n}{1 - r}\right)$

  to find the sum of the terms of the GP.

- Finally, **use the formula** $S_\infty = \dfrac{a}{1 - r}$

  to find the sum to infinity.

Q3 At the start of each year a person pays £200 into a savings account. Interest is added at a rate of 8% per year at the end of each year. Find the value of the investment after:
  (a) 2 years
  (b) 20 years.

(a) The value of the investment after 1 year is given by:

$200 \times 1.08 = £216$

and at the end of year 2 the value is:

$200 \times 1.08^2 + 200 \times 1.08 = £449.28.$

- For this type of problem it's a good idea to look at some simple cases, where there are just a few terms, to help **establish how a sequence or series can be built up**. In this question you can do this in part (a).

(b) The value of the investment after 20 years is given by:

$$200 \times 1.08 + 200 \times 1.08^2 + 200 \times 1.08^3 + \dots + 200 \times 1.08^{19} + 200 \times 1.08^{20}$$

GP with $r = 1.08$, $a = 200 \times 1.08 = 216$ and $n = 20$, which gives:

$$S_{20} = 216 \times \left(\frac{1.08^{20} - 1}{1.08 - 1}\right)$$

$$= £9884.58$$

- In part (b) it is important to **see how the terms of the progression develop and to identify the values of r, a and n** that are needed.

- Once you have these, you can substitute them into the formula

$$S_n = a\left(\frac{r^n - 1}{r - 1}\right)$$

Q4 The 8th term of an AP is 45 and the sum of the first 12 terms is 468.

(a) Find the common difference and the first term.

(b) If the sum of the first r terms is 752, find r.

(a)
$$u_n = a + (n - 1)d$$
So $45 = a + 7d$ (1)
$$S_n = \tfrac{1}{2}n(2a + (n - 1)d)$$
So $468 = 6(2a + 11d)$
$$78 = 2a + 11d \quad (2)$$
$$90 = 2a + 14d \quad 2 \times (1)$$
$$78 = 2a + 11d \quad (2)$$

Subtracting gives:
$$12 = 3d$$
$$d = 4$$

Substituting the value of d back into equation (1):
$$45 = a + 28$$
so that $a = 17$.

(b) $S_n = \tfrac{1}{2}n(2a + (n - 1)d)$

So $752 = \tfrac{1}{2}r(2 \times 17 + (r - 1) \times 4)$

$$0 = 2r^2 + 15r - 752$$

$$r = \frac{-15 \pm \sqrt{15^2 - 4 \times 2 \times (-752)}}{2 \times 2}$$

$$= 16 \text{ or } -23.5$$

$$\therefore r = 16$$

- First **use the formula for the nth term** to form one equation.

- Then **use the formula for the sum of n terms** to form a second equation, giving a pair of **simultaneous equations**.

- **Solve these** to find a value for d.

- Use this value of d to find a.

- **Use the formula for the sum of n terms** with $a = 17$, $d = 4$ and $S_r = 752$, to form a **quadratic equation**.

- **Solve the quadratic equation** to find the possible values or r.

- As r must be a **natural number**, disregard the value of −23.5.

Q1 Find the 18th term and the sum of the first 15 terms of the sequence that begins 7, 9, 11, 13, 15, …

Q2 A sequence begins: 100, 90, 81, 72.9, …
- **(a)** Find the 10th term of the sequence, correct to three decimal places.
- **(b)** Find the sum of the first 20 terms of the sequence.
- **(c)** Find the sum to infinity of the sequence.

Q3 An arithmetic series has first term 4 and common difference 3.
- **(a)** Calculate the sum of the first 20 terms.
- **(b)** The nth term of the series is greater than 103. Write down and solve an inequality, giving your answer in the form $n > a$, where a is an integer, whose value is to be found.

Q4 The 8th term of an AP is 40 and the 20th term is 124. Find the first term and the common difference. Also find the sum of the first 20 terms.

Q5 The second term of a geometric series is 80 and the fifth term of the series is 5.12.
- **(a)** Find the common ratio and the first term of the series.
- **(b)** Find the sum to infinity of the series, giving your answer as an exact fraction.
- **(c)** Find the difference between the sum to infinity of the series and the sum of the first 14 terms of the series, giving your answer in the form $a \times 10^n$, where $1 \leqslant a < 10$ and n is an integer.

Q6 The sum to infinity of a GP is exactly 7.2 and the second term of the sequence is 1. Find the two possible common ratios. For the larger of these, find the first four terms of the sequence and the sum of the first 20 terms correct to three decimal places.

Q7 The amount paid into a pension fund increases by r% each year. In the first year £1200 is paid in and in the fifth year £1932 is paid into the fund. Find r and the total paid into the fund over 10 years.

Q8 A company wants to recruit people to work for a 10-month period. They offer two different types of pay scheme.
- **A**   Starting salary of £1000 per month, increasing by £100 per month
- **B**   Starting salary of £X, increasing by 10% per month
Find X so that the total paid by both schemes is the same.

Q9 A ball rebounds to $\frac{2}{5}$ of the height from which it was dropped. The ball is dropped from a height of 2 m. Find the distance travelled by the ball, when it hits the ground for the second time. Find the total distance travelled by the ball, before it stops.

Q10 **(a)** A series is defined by $u_1 = 8$, $u_{n+1} = 0.2u_n$.
Show that the sum of the first n terms differs from the sum to infinity by an amount which is one quarter of the nth term.
- **(b)** Prove that:
$$\ln a + \ln(ab) + \ldots + \ln(ab^{n-1}) + \ldots$$
is an arithmetic series, for any positive constants a and b.
Given that the sums:
$$\ln 2 + \ln 6 + \ldots + \ln(2 \times 3^{n-1}) \quad \text{and}$$
$$\ln 3 + \ln 6 + \ldots + \ln(3 \times 2^{n-1})$$
are equal, then find n.

Answers can be found on pages 103–105.

3 Trigonometry

Key points to remember

- **Convert to and from radians** and degrees using:
  $180° = \pi$ radians
- **Arc length** $= r\theta$ when θ is in radians.
- **Area of a sector** $= \frac{1}{2}r^2\theta$ when θ is in radians.
- Remember that **trigonometric equations may have more than one solution**.
- **Know how to find the solutions in a given range**. For example the equation $\sin 2x = 0.5$ has four solutions in the interval $0 \leq x \leq 360°$, as shown on the graph.

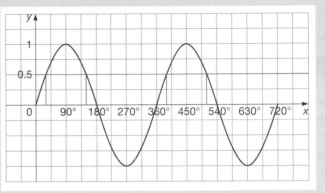

- The graph below shows $y = A\sin(\omega x + \alpha) + d$.

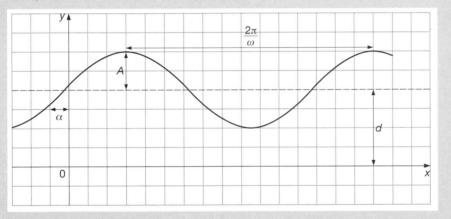

- **Pythagorean identities**
  $$\cos^2\theta + \sin^2\theta = 1$$
  $$1 + \tan^2\theta = \sec^2\theta$$
  $$\cot^2\theta + 1 = \operatorname{cosec}^2\theta$$

- **Sine rule**
  $$\frac{a}{\sin A} = \frac{b}{\sin B} = \frac{c}{\sin C}$$

- **Cosine rule**
  $$a^2 = b^2 + c^2 - 2bc\cos A$$

- Know that:
  $$\tan\theta = \frac{\sin\theta}{\cos\theta}$$

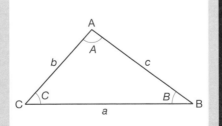

Formulae you must know

- Arc length $= r\theta$
- Area of a sector $= \frac{1}{2}r^2\theta$
- $\cos^2\theta + \sin^2\theta = 1$
- $1 + \tan^2\theta = \sec^2\theta$
- $\cot^2\theta + 1 = \operatorname{cosec}^2\theta$
- $\tan\theta = \dfrac{\sin\theta}{\cos\theta}$
- $\dfrac{a}{\sin A} = \dfrac{b}{\sin B} = \dfrac{c}{\sin C}$
- $a^2 = b^2 + c^2 - 2bc\cos A$

Don't forget to **include all the solutions** when solving an equation.

Don't **mix up radians and degrees**.

Don't make **arithmetic or algebraic errors** when rearranging equations.

Exam Questions and Student's Answers

How to score full marks

Q1 The diagram shows a circle of radius 4 cm and centre O. The two points, A and B, lie on the circle such that

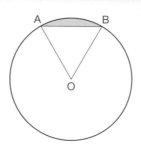

$\angle AOB = \dfrac{\pi}{6}$ radians.

Find the area of the shaded region.

Area of sector = $\dfrac{1}{2} \times 4^2 \times \dfrac{\pi}{6}$

$= \dfrac{4}{3}\pi \, cm^2$

Area of triangle = $\dfrac{1}{2} \times 4 \times 4\sin\dfrac{\pi}{6}$

$= 4 \, cm^2$

Area of shaded region

$= \dfrac{4}{3}\pi - 4$

$= 0.189 \, cm^2$ to three decimal places.

- **First find the area of the sector** using the formula $A = \frac{1}{2}r^2\theta$.

- **To find the area of the triangle, draw in the height of the triangle**, form an expression for the height and then calculate the area of the triangle.

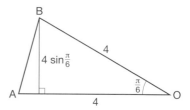

- The **area of the shaded region** is the area of the sector *minus* the area of the triangle.

Q2 Solve the equation $4\sin x = 3$, giving all the solutions in the range $0 \leqslant x \leqslant 360°$.

$4\sin x = 3$

$\sin x = \dfrac{3}{4}$

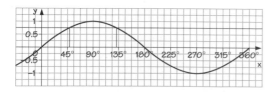

$x = 48.6°$ or $x = 180° - 48.6° = 131.4°$

$x = 48.6°$ or $x = 131.4°$

- First **rewrite the equation in the form** $\sin x = \dots$.

- The graph shows that **there are two solutions** in the range $0 \leqslant x \leqslant 360°$.

- If you use a calculator you find $x = 48.6°$ (correct to one decimal place) as the first solution. From the graph you can see that this value must be subtracted from $180°$ to get the second solution.

Q3 Solve the equation $4\cos 2x + 1 = 3$, giving all the solutions in the range $0 \leqslant x \leqslant 360°$.

$4 \cos 2x + 1 = 3$

So $\cos 2x = \dfrac{1}{2}$

This gives $2x = 60°$. Other values of $2x$ that satisfy this equation can be found from the graph of $y = \cos x$.

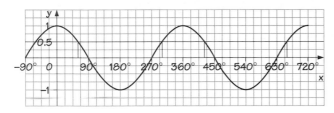

The other solutions are:

$2x = 360° - 60° = 300°$,
$2x = 360° + 60° = 420°$,
$2x = 720° - 60° = 660°$

$60° \div 2 = 30°$, $300° \div 2 = 150°$,
$420° \div 2 = 210°$, $660° \div 2 = 330°$
$x = 30°, 150°, 210°, 330°$

Q4 Solve the equation $5 = 6\cos^2\theta + \sin\theta$, giving all the solutions in the range $0 \leqslant \theta \leqslant 360°$.

$5 = 6\cos^2\theta + \sin\theta$
$5 = 6(1 - \sin^2\theta) + \sin\theta$
$6\sin^2\theta - \sin\theta - 1 = 0$
$(2\sin\theta - 1)(3\sin\theta + 1) = 0$
$\sin\theta = \dfrac{1}{2}$ and $\sin\theta = -\dfrac{1}{3}$

For $\sin\theta = \dfrac{1}{2}$, the solutions are $\theta = 30°$

and $\theta = 180° - 30° = 150°$.

For $\sin\theta = -\dfrac{1}{3}$, a calculator will give $\theta = -19.5°$. The solutions in the range $0 \leqslant \theta \leqslant 360°$ are
$\theta = 180° + 19.5° = 199.5°$ and
$\theta = 360° - 19.5° = 340.5°$.

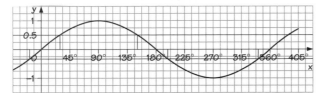

How to score full marks

- Once you have written the equation in the form $\cos 2x = \ldots$ then you can find a value for $2x$. **You should recognise that $2x = 60°$ in this case, but at other times you may need to use a calculator.**

- As **the equation contains $2x$**, you first need to **find all the solutions between 0 and 720°, so that when they are halved you obtain solutions in the range 0 to 360°.**

- The graph shows the solution for $2x$.

- These are the possible values of $2x$. The corresponding values of x can now be found by dividing by 2.

- **Using the identity $\cos^2\theta + \sin^2\theta = 1$, in the form $\cos^2\theta = 1 - \sin^2\theta$, you can eliminate the $\cos^2\theta$ term from the equation, to produce a quadratic equation.**

- In this problem it is easy to **factorise the equation**, but in other cases you may need to use the **quadratic equation formula**.

- As there are **two solutions** to the quadratic equation there are **four values for θ**, as shown on the graph.

- Note that when you use $\sin^{-1}$ on your calculator for negative values, you will get a negative answer. **You need to be able to calculate angles in the range 0 to 360°.**

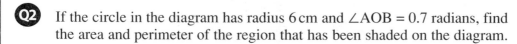

Q1 A and B are points on a circle, centre O, of radius r cm. The length of the arc AB is $2r$ cm. Find the area of the sector AOB in terms of r.

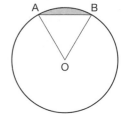

Q2 If the circle in the diagram has radius 6 cm and $\angle AOB = 0.7$ radians, find the area and perimeter of the region that has been shaded on the diagram.

Q3 A manufacturer needs to make a thin metal plate in the shape of a circular sector, with perimeter 20 cm. The diagram shows such a sector with radius r cm, angle θ radians and area A cm^2.

(a) Prove that $A = 25 - (r - 5)^2$.

Given that r can vary:

(b) deduce the value of r for which A is a maximum and state the maximum value of A.

Q4 Find the values of θ between 0 and 360° for which $\sin\theta = -0.5$.

Q5 Solve the equation $2 - 4\sin 2x = 0$, giving all the solutions in the range $0 \leqslant x \leqslant 360°$.

Q6 Find, in radians, two angles θ, $0 < \theta < 2\pi$, for which $3\tan\theta = \dfrac{1}{\cos\theta}$. Give each answer to three places of decimals.

Q7 Find all solutions of the equation $2\sin^2 x + 5\cos x = 0$ in the interval $0 \leqslant x \leqslant 360°$, giving your answer to the nearest 0.1°.

Q8 Show that the equation $15\cos^2\theta = 13 + \sin\theta$ may be written as a quadratic equation in $\sin\theta$.

Hence solve the equation, giving all values of θ such that $0 \leqslant \theta \leqslant 360°$.

Q9 Find, in degrees, the values of θ in the interval $0 \leqslant \theta \leqslant 360°$ for which

$$4\sin^2\theta - 2\sin\theta = 4\cos^2\theta - 1.$$

State which of your answers are exact and which are given to a degree of accuracy of your choice, which you should give.

Q10 The depth of water at the entrance to a harbour is y metres at time t hours after low tide. The value of y is given by $y = 10 - 3\cos kt$, where k is a positive constant.

(a) Write down or obtain, the depth of water in the harbour:
    (i) at low tide
    (ii) at high tide.
(b) Show by means of a sketch graph how y varies with t between two successive low tides.
(c) Given that the time interval between a low tide and the next high tide is 6.20 hours, calculate, correct to two decimal places, the value of k.

Answers can be found on pages 105–106.

Key points to remember

- The **equation of a straight line** can be written as $y = mx + c$, where m is the gradient and c is the intercept with the vertical axis.

- The **straight line through two points** with coordinates (x_1, y_1) and (x_2, y_2) has gradient $\dfrac{y_2 - y_1}{x_2 - x_1}$ and equation $\dfrac{y - y_1}{y_2 - y_1} = \dfrac{x - x_1}{x_2 - x_1}$.

- If the **gradient of a line is m**, then the **gradient of a line perpendicular to it is** $-\dfrac{1}{m}$.

- The **product of the gradients of two perpendicular lines** is -1.

- The **distance between the points** with coordinates (x_1, y_1) and (x_2, y_2) is $\sqrt{(x_2 - x_1)^2 + (y_2 - y_1)^2}$.

- The **midpoint of the line joining two points** with coordinates (x_1, y_1) and (x_2, y_2) is $\left(\dfrac{x_1 + x_2}{2}, \dfrac{y_1 + y_2}{2}\right)$.

Formulae you must know

- $y = mx + c$

- $m = \dfrac{y_2 - y_1}{x_2 - x_1}$

- $\dfrac{y - y_1}{y_2 - y_1} = \dfrac{x - x_1}{x_2 - x_1}$

- $\sqrt{(x_2 - x_1)^2 + (y_2 - y_1)^2}$

- $\left(\dfrac{x_1 + x_2}{2}, \dfrac{y_1 + y_2}{2}\right)$

BRUCE'S TEACHER HAD SET A PROJECT ON GRADIENTS ...

Don't make these mistakes ...

Don't omit the negative sign when finding the gradient of a perpendicular to a line.

Don't make silly mistakes when using **negative coordinates** to find gradients.

Don't forget to **calculate the y-coordinate when you have found the x-coordinate.**

Q1 The points A and B have coordinates (7, 1) and (−1, −3) respectively. The point C is such that ABC is a right-angled triangle, with the right angle at C. The side AC is parallel to the line $3y + x − 12 = 0$. Find the coordinates of the point C.

$3y + x − 12 = 0$

$y = −\frac{1}{3}x + 4$ so it has gradient $−\frac{1}{3}$.

The line through A and C has equation $y = −\frac{1}{3}x + c$. Using $x = 7$ and $y = 1$ gives:

$1 = −\frac{7}{3} + c \Rightarrow c = \frac{10}{3}$

$y = −\frac{1}{3}x + \frac{10}{3}$

A line through the points B and C has gradient 3. Its equation will be $y = 3x + c$. Using $x = −1$ and $y = −3$ gives:

$−3 = −3 + c \Rightarrow c = 0$

So the equation of BC is $y = 3x$.

The lines intersect when:

$3x = −\frac{1}{3}x + \frac{10}{3}$ and $y = 3 \times 1$

  $x = 1$ $y = 3$

So the point of intersection is at (1, 3).

- **Use the information about the parallel line** to find the **gradient of the line** AC.

- You can use the **coordinates of the point A** to find the value of c in the equation.

- As the lines AC and BC are perpendicular, the **product of their gradients must be −1**, so:

  $$\text{gradient of BC} = \frac{−1}{\text{gradient of AC}}$$

- When the lines intersect, they will have the same y-value. Use this to **form and solve an equation** to find x.

Q2 The points A and B have coordinates (4, 2) and (6, 8) respectively. Find the equation of the line perpendicular to AB that passes through its midpoint.

The midpoint of AB has coordinates

$\left(\dfrac{4 + 6}{2}, \dfrac{2 + 8}{2}\right) = (5, 5)$.

The gradient of AB is $\dfrac{8 − 2}{6 − 4} = 3$.

The equation of the line is $y = −\frac{1}{3}x + c$.

When $x = 5$, $y = 5$.

So $5 = −\frac{5}{3} + c \Rightarrow c = \frac{20}{3}$

and $y = −\frac{1}{3}x + \frac{20}{3}$.

- **Find the midpoint and gradient of the line AB**.

- If **the gradient of the line AB is m**, then **the gradient of the line perpendicular to it will be $−\dfrac{1}{m}$**, in this case $−\frac{1}{3}$.

- Use the **coordinates of the midpoint** to find the value of the constant c.

Questions to try

Q1 The points A and B have coordinates (1, 6) and (5, 14) respectively. Find the equation of the line *p* that passes through the points A and B, and the line *q* that is perpendicular to *p* and passes through the midpoint of AB.

Q2 The lines $y - 3x + 6 = 0$ and $3y + x - 12 = 0$ intersect at the point A.

   (a) Find the coordinates of A.
   (b) Show that the two lines are perpendicular.
   (c) Find the area of the triangle formed by the two lines and the *x*-axis.

Q3 The line *p* is parallel to the line with equation $3y - x + 4 = 0$, and intersects the *x*-axis at (–2, 0). The line *q* is perpendicular to the line *p* and intersects the *x*-axis at (8, 0). Find the equation of each line and the coordinates of their point of intersection.

Q4 The points A, B and C have coordinates (3, 6), (6, 5) and (7, 2) respectively.

   (a) Find the equation of the line *p* that passes through the points A and C and the line *q* that is perpendicular to *p* and passes through the point B.
   (b) Find the coordinates of the point where the lines *p* and *q* intersect.

Q5 The points A(–2, 4), B(6, –2) and C(5, 5) are the vertices of $\triangle ABC$ and D is the midpoint of AB.

   (a) Find an equation of the line passing through A and B in the form $ax + by + c = 0$, where *a*, *b* and *c* are integers to be found.
   (b) Show that CD is perpendicular to AB.

Answers can be found on pages 106–107.

Key points to remember

- A function is a one-to-one or many-to-one mapping.
- A function maps members of its domain onto members of its range
- If f is a function, its inverse is f^{-1}. When f is one-to-one then f^{-1} exists. To find f^{-1}, solve f(x) = y for x.

- Composite functions

  fg(x) = f(g(x))

- Formulae you must know
  - $|x| = x$ if $x \geq 0$ or $-x$ if $x < 0$

Don't confuse the **order of composite functions**.

Don't leave **inverse functions** in terms of y.

Don't forget that **some equations that you have to solve may have more than one solution**.

Don't forget to **state the domain of composite and inverse functions**.

Don't make **algebraic errors** when simplifying **composite functions** or **finding inverses**.

Exam Questions and Student's Answers

How to score full marks

Q1 The functions f and g are given by:

f: $x \mapsto x + 3, x \in \mathbb{R}$
g: $x \mapsto x^2 - 1, x \in \mathbb{R}$

(a) State the range of g.
(b) Find gf(2).
(c) Find gf(x).
(d) Solve fg(x) = 11.

(a) The range is g(x) $\geqslant$ –1, as $x^2 \geqslant 0$.

(b) gf(2) = g(f(2))

= g(2 + 3)

= g(5)

= $5^2 - 1$

= 24

(c) gf(x) = g(f(x))

= g(x + 3)

= $(x + 3)^2 - 1$

= $x^2 + 6x + 8$

(d) fg(x) = f(g(x))

= f($x^2 - 1$)

= $x^2 + 2$

fg(x) = 11

$x^2 + 2 = 11$

$x^2 = 9$

$x = \pm 3$

- **You can find the range** by considering all the possible values that could be produced by the function. As $x^2 \geqslant 0$ for all values of x, the range can be found.

- **You can find gf(2)** by first finding f(2) and then applying g to the result.

- **To find gf(x)** by taking f(x), which is given, and applying the function g to this.

- **First find fg(x)** – remember to apply the function f to g(x).

- Then **form the equation fg(x) = 11** and solve this to find the values of x.

Q2 The functions f and g are given by:

f: $x \mapsto e^x + 2, x \in \mathbb{R}$
g: $x \mapsto 2x - 1, x \in \mathbb{R}$

(a) State the range of f.
(b) Find $g^{-1}(x)$ and $f^{-1}(x)$.

(a) The range of f is f(x) > 2.
(b) First find $g^{-1}(x)$.

$y = 2x - 1$

$y + 1 = 2x$

$x = \dfrac{y + 1}{2}$

Then $g^{-1} : x \mapsto \dfrac{x + 1}{2}, x \in \mathbb{R}$.

Then find $f^{-1}(x)$.

$y = e^x + 2$

$e^x = y - 2$

$x = \ln(y - 2)$

Then $f^{-1} : x \mapsto \ln(x - 2), x \in \mathbb{R}, x > 2$

- You can find the range by **considering all the possible values that could be produced by the function**. As $e^x > 0$ for all values of x, the range can be found.

- **To find the inverse**, $g^{-1}(x)$, write $y = g(x)$ and solve for x.

- Then replace y by x and write as a function. **Note that the domain contains all values of x.**

- Repeat the process for the function $f(x)$, **noting the use of logarithms**.

- The domain is $x > 2$, because the range of f(x) is f(x) > 2.

Q3 If $f(x) = 2x - 4$:
(a) sketch the graph of $y = |f(x)|$
(b) solve the equation $x = |f(x)|$.

f(x) > 0 if x > 2 and f(x) < 0 if x < 2. Also f(x) = 0 when x = 2.

|f(x)| = 2x − 4 if x ⩾ 2
or 4 − 2x if x < 2.

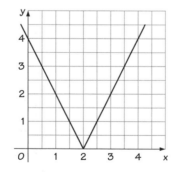

- You should note that because $f(x)$ is a **linear function** the graph will consist of **straight lines**.

- The **function will be defined in two different ways**, either side of $x = 2$.

- The **important points** to indicate **on the graph** are where the **function crosses the axes**.

(b) If x ⩾ 2 then

x = 2x − 4

x = 4

If x < 2 then

x = 4 − 2x

3x = 4

$x = \dfrac{4}{3}$

- You must consider both the cases, where $x \geqslant 2$ and $x < 2$, because these will produce two different equations and hence two different values of x.

Q1 The functions f and g are given by:

f: $x \mapsto x - 3$, $x \in \mathbb{R}$
g: $x \mapsto 4x^2$, $x \in \mathbb{R}$

(a) State the range of g.
(b) Find gf(4) and fg(1).
(c) Find gf(x) and fg(x). State the range of each of these composite functions.
(d) Find f^{-1}.

Q2 The functions f and g are given by:

f: $x \mapsto e^x$, $x \in \mathbb{R}$
g: $x \mapsto x + 1$, $x \in \mathbb{R}$

(a) State the range of f.
(b) Find gf(x) and fg(x). State the range of each of these composite functions.
(c) Find f^{-1} and g^{-1}, stating the domain of each function.

Q3 The function f is given by:

f: $x \mapsto \dfrac{x + 6}{x}$, $x \in \mathbb{R}$, $x \neq 0$.

(a) Find f^{-1}.
(b) Find the values of x that satisfy $f(x) = f^{-1}(x)$.

Q4 The functions f and g are given by:

f: $x \mapsto x - 1$, $x \in \mathbb{R}$
g: $x \mapsto x^2 - 1$, $x \in \mathbb{R}$

(a) Sketch the graph of $y = |f(x)|$ and solve the equation $|f(x)| = 7$.
(b) Find gf(x) and sketch the graphs of $y = g(x)$ and $y = gf(x)$.
(c) Find the values of x for which $|gf(x)| = 1$.

Q5 The functions f and g are defined as:

f: $x \mapsto x^2$, $x \in \mathbb{R}$
g: $x \mapsto a - x$, $x \in \mathbb{R}$

where a is a positive constant.

(a) Find fg(x) and $\text{gf}\left(\dfrac{\sqrt{a}}{2}\right)$.
(b) Sketch the graph of $y = |g(x)|$ and find the values of x that satisfy $|g(x)| = 2$.

Q6 The functions f and g are given by:

f: $x \mapsto x + 2$, $x \in \mathbb{R}$
g: $x \mapsto e^x$, $x \in \mathbb{R}$

(a) Find fg(x) and gf(x).
(b) If $h(x) = gf(x)$, is it true that $f^{-1}g^{-1}(x) = h^{-1}(x)$?

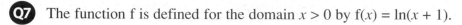

Q7 The function f is defined for the domain $x > 0$ by $f(x) = \ln(x + 1)$.

    (a) Find the range of f.
    (b) Find an expression for $f^{-1}(x)$ in terms of x and e.

Q8 The functions f and g are defined with their respective domains by:

$$f: x \mapsto \frac{5}{x - 2}, x \in \mathbb{R}, x \neq 2$$

$$g: x \mapsto x^2 + 3, x \in \mathbb{R}$$

    (a) Calculate the exact values of x for which $f(x) = x$.
    (b) Find the range of g.
    (c) The domain of the composite function fg is $\mathbb{R}$. Find $fg(x)$ and state the range of fg.
    (d) State whether the inverse of g exists, giving reason for your answer.

Q9 The functions f and g are defined by:

$$f: x \mapsto 1 + x^{\frac{1}{2}}, x \geq 0$$
$$g: x \mapsto x^2, x \in \mathbb{R}$$

    (a) Find the domain of the inverse function f^{-1}.
    (b) Find an expression for $f^{-1}(x)$.
    (c) Find and simplify an expression for $fg(x)$ for the case where $x \geq 0$.
    (d) Explain clearly why the value of $fg(-2)$ is 3.
    (e) Sketch the graph of $y = fg(x)$, for both positive and negative values of x, and give the equation of this graph in a simplified form.

Q10 The function f is given by:

$$f: x \mapsto \ln(4 - 2x), x \in \mathbb{R}, x < 2$$

    (a) Find an expression for $f^{-1}(x)$.
    (b) Sketch the curve with equation $y = f^{-1}(x)$, showing the coordinates of the points where the curve meets the axes.

The function g is given by:

$$g: x \mapsto 3^x, x \in \mathbb{R}$$

    (c) Find the value of x for which $g(x) = 1.5$, giving your answer to three decimal places.
    (d) Evaluate $gf(1)$ to three decimal places.

Answers can be found on pages 107–109.

Key points to remember

• Techniques

Simplify brackets and fractions before differentiating.

- Write $(x^2 - 4)x$ as $x^3 - 4x$ before differentiating.
- Write $\dfrac{x^2 + 2}{x}$ as $x + 2x^{-1}$ before differentiating.

Write reciprocals and roots as powers.

- Write $\sqrt{x}$ as $x^{\frac{1}{2}}$ and $\dfrac{1}{x^3}$ as x^{-3} before differentiating.

• Gradients of curves

The **gradient of a curve** is given by its **derivative**.

• Increasing and decreasing functions

A function $f(x)$ is **increasing** on the interval (a, b), if $f'(x) > 0$, for all x in (a, b).

A function $f(x)$ is **decreasing** on the interval (a, b), if $f'(x) < 0$, for all x in (a, b).

• Stationary points

At a stationary point $\dfrac{dy}{dx} = 0$.

A stationary point can be a **local maximum**, a **local minimum** or a **point of inflexion**. Test by considering the gradient either side of the stationary point.

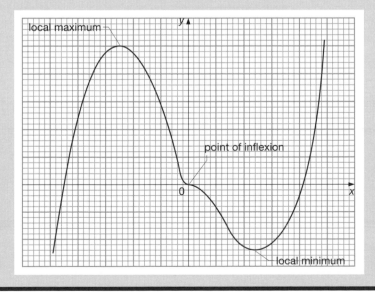

I THOUGHT I'D REACHED THE TOP—BUT IT'S ONLY A LOCAL MAXIMUM!

Formulae you must know

- Derivative of x^n

$$\frac{d}{dx}(x^n) = nx^{n-1}$$

- Function notation

If $y = f(x)$ then $\dfrac{dy}{dx} = f'(x)$

Don't make these mistakes...

Don't forget to **simplify brackets or fractions** before differentiating.

Don't forget to **check what type of stationary point** you have found.

Don't forget to **use the rules of indices**, where you need them.

Don't make silly mistakes when **subtracting 1** from a fraction.

Exam Questions and Student's Answers

Q1 If $y = x^2\left(3x - 1 + \dfrac{1}{x}\right)$ find $\dfrac{dy}{dx}$.

$$y = x^2\left(3x - 1 + \frac{1}{x}\right) = 3x^3 - x^2 + x$$

So $\dfrac{dy}{dx} = 9x^2 - 2x + 1$

Q2 If $f(x) = \dfrac{\sqrt{x} + 2}{x^2}$ find $f'(x)$.

$$f(x) = \frac{\sqrt{x} + 2}{x^2}$$

$$= \frac{\sqrt{x}}{x^2} + \frac{2}{x^2}$$

$$= \frac{x^{\frac{1}{2}}}{x^2} + \frac{2}{x^2}$$

$$= x^{-\frac{3}{2}} + 2x^{-2}$$

So $f'(x) = -\dfrac{3}{2}x^{-\frac{3}{2}-1} - 4x^{-2-1}$

$$= -\frac{3}{2}x^{-\frac{5}{2}} - 4x^{-3}$$

$$= -\frac{3}{2\sqrt{x^5}} - \frac{4}{x^3}$$

How to score full marks

- **Expand the brackets** before differentiating.

- Remember to **multiply each term inside the bracket** by x^2.

- You must **simplify** each term before differentiating.

- Differentiate each term using the rule for x^n.

- You don't know how to **differentiate a fraction** like this, so you must **simplify it** first.

- First **split the fraction into two parts**.

- Then write them in the form x^n.

- Now you can differentiate using **the rule for differentiating x^n**.

- You can also write the result in this form.

Q3 If $f(x) = (x - 3)^2 + 1$ find the values of x for which $f(x)$ is increasing.

$f(x) = (x - 3)^2 + 1$

$\quad = x^2 - 6x + 9 + 1$

$\quad = x^2 - 6x + 10$

$f'(x) = 2x - 6$

If $f'(x) > 0$ then:

$2x - 6 > 0$

$\quad 2x > 6$

$\quad\ x > 3$

How to score full marks

- For an **increasing function** you need $f'(x) > 0$.

- First **expand the brackets**. Then differentiate to find $f'(x)$.

- Use the **derivative to form an inequality**. (Don't start with an equation.)

- **Solve the inequality** to find the solution.

Q4 Determine the coordinates of the stationary points of the curve with equation $y = \dfrac{2}{x} + 8x$.

Also determine whether each of these points is a local maximum or minimum.

$y = 2x^{-1} + 8x$

$\dfrac{dy}{dx} = -2x^{-2} + 8$

For stationary points $\dfrac{dy}{dx} = 0$.

$0 = -\dfrac{2}{x^2} + 8$

$x^2 = \dfrac{1}{4}$

$x = \dfrac{1}{2}$ or $x = -\dfrac{1}{2}$

x	$\frac{1}{4}$	$\frac{1}{2}$	1
$\dfrac{dy}{dx}$	-24	0	6

There is a local minimum at $(\frac{1}{2}, 8)$.

x	-1	$-\frac{1}{2}$	$-\frac{1}{4}$
$\dfrac{dy}{dx}$	6	0	-24

There is a local maximum at $(-\frac{1}{2}, -8)$.

- First **write $\dfrac{2}{x}$ as $2x^{-1}$**, so that it can be **differentiated**.

- Remember that **at stationary points the derivative is zero**.

- Solve the equation that you form, **finding both values of x**.

- **Substitute values for x either side of $x = \frac{1}{2}$ to find the gradients**. Here the gradient changes from being negative to positive at the stationary point so it is a local minimum. **Don't forget to calculate the y-coordinate as well**.

- For the negative value of x, the gradient changes from being positive to negative, indicating that there is a **local maximum**.

Q1 Given that $y = x^4 - 3x^2 - x - 2$, find $\dfrac{dy}{dx}$.

Q2 If $y = x^3 + 6x - 8$ find $\dfrac{dy}{dx}$.

Q3 If $f(x) = \sqrt{x}\left(x + \dfrac{2}{x}\right)$, find $f'(x)$.

Q4 If $f(x) = x^2 + \dfrac{16}{x}$, find the range of positive values of x for which $f(x)$ is increasing.

Q5 A curve has equation $y = (x^2 + 3)(2x - 1)$.

  (a) Find $\dfrac{dy}{dx}$.

  (b) Determine the coordinates of the points on the curve where the gradient is equal to 6.

Q6 The equation of a curve is $y = 9x^2 - 4x^3$. Find the coordinates of the two stationary points on the curve, and determine the nature of these stationary points.

Determine the set of values of x for which $9x^2 - 4x^3$ is a decreasing function of x.

Q7 A rectangular paddock with sides of length x and y is to be built using 20 m of fencing. One side of the rectangle will be against a stone wall and so will not need to be fenced. The diagram shows the rectangle.

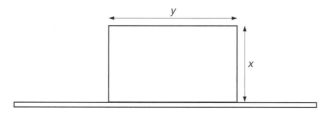

Show that the area of the rectangle, A cm^2, is given by $A = 20x - 2x^2$ and find the value of x for which the area is a maximum.

Q8 A farmer wishes to construct a rectangular cattle pen. He has 63 m of fencing available and must leave a gap of 5 m in the middle of one side of the rectangle to allow cattle to enter the pen.

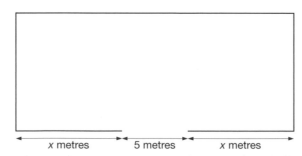

He constructs the pen so that there are x metres of fencing each side of the gap, as shown in the diagram. He uses all 63 m of the fencing available. The area enclosed is $A\,\text{m}^2$.

(a) Show that $A = (2x + 5)(29 - 2x)$.

(b) Find the maximum value of A.

 Q9 An architect is drawing up plans for a mini-theatre. The diagram shows the plan of the base which consists of a rectangle of length $2y$ metres and width $2x$ metres and a semicircle of radius x metres which is placed with one side of the rectangle as diameter.

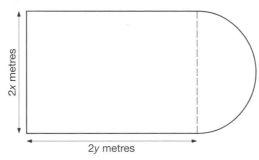

2x metres

2y metres

Find, in terms of x and y, expressions for:

(a) the perimeter of the base

(b) the area of the base.

The architect decides that the base should have a perimeter of 100 metres.

(c) Show that the area, A square metres, of the base is given by:

$$A = 100x - 2x^2 - \tfrac{1}{2}\pi x^2$$

(d) Given that x can vary, find the value of x for which $\dfrac{\text{d}A}{\text{d}x} = 0$ and determine

the corresponding value of y, giving your answer to two significant figures.

(e) Find the maximum value of A and explain why this is a maximum.

Q10 An open-topped rectangular box is to have a volume of $500\,\text{cm}^3$. The box has a square base with sides of length $x\,\text{cm}$. Show that the external surface area of the box, $S\,\text{cm}^2$, is given by:

$$S = x^2 + \frac{2000}{x}$$

and find the dimensions of the box for which the surface area is a minimum.

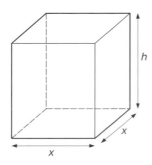

h

x

x

Answers can be found on pages 109–110.

Key points to remember

● Areas under curves

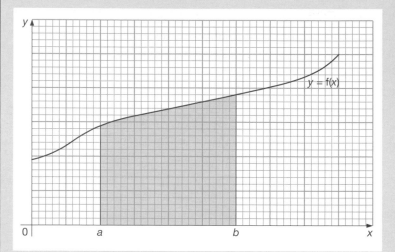

Area of shaded region = $\int_a^b f(x)dx$.

Where an integral is used to find an area under the x-axis, the result will be **negative**.

For example consider the curve $y = x^2 - 4$, as shown in the diagram.

$\int_{-2}^2 (x^2 - 4)dx = -\frac{32}{3}$

$\int_2^4 (x^2 - 4)dx = \frac{32}{3}$

$\int_{-2}^4 (x^2 - 4)dx = 0$

But the area of the shaded region is

$\frac{32}{3} + \frac{32}{3} = \frac{64}{3}$

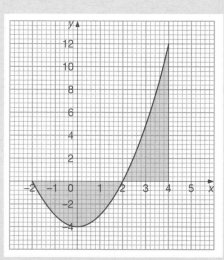

When finding an area such as the one shaded in this diagram, you must **evaluate each part separately** and change the sign on any negative parts before calculating the total area.

● Definite integrals

These are integrals such as $\int_0^2 x^2 dx$, which are evaluated to give a **specific value**.

● Indefinite integrals

These are integrals such as $\int x^2 dx$, which will be in terms of x and must include a **constant of integration**, *c*.

THE AREA UNDER THIS CURVE IS POSITIVELY DRY!

Formulae you must know

● Standard integrals

$\int x^n dx = \frac{x^{n+1}}{n+1} + c$ if $n \neq -1$

$\int \frac{1}{x} dx = \ln|x| + c$

Don't make these mistakes...

Don't make silly mistakes when **adding 1 to fractions or negative numbers**.

Don't forget that **curves that are below the x-axis will produce negative values** when integrated.

Don't forget to **include the constant of integration** for indefinite integrals.

Don't use the wrong **limits of integration**.

Don't forget to **divide by the new power**, when integrating.

Don't make errors when **dividing by a fraction**.

Exam Questions and Student's Answers

How to score full marks

Q1 Find $\int \sqrt{x}\left(x + \frac{1}{x}\right) dx$

$$\int \sqrt{x}\left(x + \frac{1}{x}\right) dx = \int \left(x\sqrt{x} + \frac{1}{\sqrt{x}}\right) dx$$

$$= \int \left(x^{\frac{3}{2}} + x^{-\frac{1}{2}}\right) dx$$

$$= \frac{2}{5}x^{\frac{5}{2}} + 2x^{\frac{1}{2}} + c$$

- First **expand the bracket** and write each term in the form x^n, **then integrate**.

- Note that as this is an **indefinite integral**, you need **a constant of integration**.

Q2 Find the **area enclosed by the curve** $y = \frac{1}{x}$, the x-axis and the lines $x = 1$ and $x = 4$.

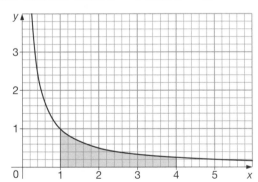

The diagram shows the area that was asked for.

The area is given by:

$$\text{area} = \int_1^4 \frac{1}{x} dx$$

$$= \left[\ln|x|\right]_1^4$$

$$= \ln 4 - \ln 1$$

$$= \ln 4$$

- Drawing **a sketch of the area required helps you to identify the limits of integration**.

- After integrating, **first substitute the upper limit and then the lower limit**. Note that **you don't need to introduce a constant of integration for a definite integral**.

- Note that $\ln 1 = 0$.

Q3 Find the area of the region bounded by the curves $y = \dfrac{1}{x}$ and $y = 8x^2$, the x-axis and the line $x = 2$.

- The first stage is to **find where the two curves intersect**. At this point both curves will have the same y-coordinate, and you need to use this to form the equation in the solution.

$8x^2 = \dfrac{1}{x}$

$x^3 = \dfrac{1}{8}$

$x = \dfrac{1}{2}$

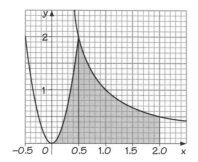

- The diagram shows the area that was asked for.

The two curves intersect when $x = \dfrac{1}{2}$.

So the area is given by:

$$\int_0^{\frac{1}{2}} 8x^2 \, dx + \int_{\frac{1}{2}}^{2} \dfrac{1}{x} \, dx = \left[\dfrac{8x^3}{3}\right]_0^{\frac{1}{2}} + \Big[\ln|x|\Big]_{\frac{1}{2}}^{2}$$

- To find the total area, you must **evaluate two integrals**: one from 0 to 0.5 for the first curve and the other from 0.5 to 2 for the second curve.

$$= \dfrac{1}{3} - 0 + \ln 2 - \ln \dfrac{1}{2}$$

$$= \dfrac{1}{3} + 2\ln 2$$

Q4 The points A and B lie on the curve $y = 5 + 2x - x^2$ and have coordinates $(0, 5)$ and $(3, 2)$ respectively. Find the area of the region enclosed by the curve and the line AB.

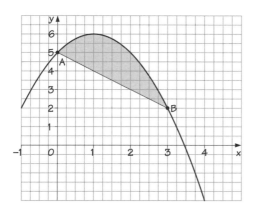

- It **is important** that you draw a diagram to **identify the area** that is required.

- The shaded area is given by $\int_0^3 (5 + 2x - x^2) \, dx$ less the area of the trapezium with corners at $(0, 0)$, $(3, 0)$, $(3, 2)$ and $(0, 5)$. So you can evaluate the area as shown.

$$\int_0^3 (5 + 2x - x^2) \, dx - \dfrac{1}{2} \times 3(5 + 2)$$

$$= \left[5x + x^2 - \dfrac{x^3}{3}\right]_0^3 - \dfrac{21}{2}$$

$$= 15 + 9 - \dfrac{27}{3} - \dfrac{21}{2}$$

$$= \dfrac{9}{2}$$

- Note the use of the formula $A = \dfrac{1}{2}h(a + b)$ for the area of the **trapezium**.

Q1 Find $\int_0^3 (2x^3 - x^2)\,\mathrm{d}x$. **Q2** Find $\int \dfrac{(x^2 - x)}{\sqrt{x}}\,\mathrm{d}x$. **Q3** Find $\int \dfrac{1}{x^2}(\sqrt{x} - x)\,\mathrm{d}x$.

Q4 Find the area of the region bounded by the curve $y = x^2$, the line $y = 12 - x$ and the x-axis.

Q5 Find the area of the region bounded by the curve $y = x^2 - 5x + 9$ and the line $y = 3$.

Q6 The points A and B lie on the curve $y = 16 - x^4$ and have coordinates $(-2, 0)$ and $(1, 15)$ respectively. Find the area of the finite region enclosed by the curve and the line AB.

Q7 The points A and B lie on the curve $y = x^2(5 - x)$ and have coordinates $(1, 4)$ and $(4, 16)$ respectively. Find the area of the region enclosed by the curve and the chord AB.

Q8 The graph on the right shows the curve with equation $y = x^3 - 5x^2 + 6x$. Find the total area of the regions that have been shaded on the diagram.

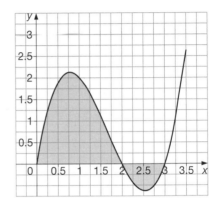

Q9 (a) Evaluate $\int_0^4 x(4 - x)\,\mathrm{d}x$.

(b) The diagram on the right shows the curve $y = x(4 - x)$, together with a straight line. The line cuts the curve at the origin O and at the point P with x-coordinate k, where $0 < k < 4$.

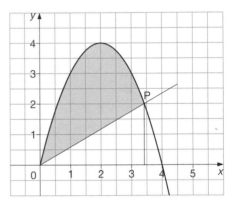

  (i) Show that the area of the shaded region, bounded by the curve and the line, is $\frac{1}{6}k^3$.

  (ii) Find, correct to three decimal places, the value of k for which the area of the shaded region is half of the total area under the curve between $x = 0$ and $x = 4$.

Q10 The diagram on the right shows sketches of the line with equation $x + y = 4$ and the curve with equation $y = x^2 - 2x + 2$ intersecting at points P and Q. The minimum point on the curve is M. The shaded region R is bounded by the line and the curve.

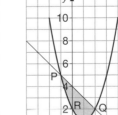

(a) Show that the coordinates of M are $(1, 1)$.
(b) Find the coordinates of P and Q.
(c) Prove that the triangle PMQ is right-angled and hence show that the area of the triangle PMQ is 3 square units.
(d) Show that the area of the region R is $1\frac{1}{2}$ times that of the area of the triangle PMQ.

Answers can be found on pages 110–112.

Key points to remember

● The factor theorem

For any polynomial function $f(x)$:

$f(a) = 0 \Leftrightarrow (x - a)$ is factor.

More generally:

$f\left(\dfrac{a}{b}\right) = 0 \Leftrightarrow (bx - a)$ is a factor.

● Fixed point iteration

To solve an equation $f(x) = 0$, rearrange it into the form $x = g(x)$.

The iteration $x_{n+1} = g(x_n)$ for $n = 0, 1, 2, \ldots$

produces a **convergent sequence** towards a root α provided that $|g'(\alpha)| < 1$ and the initial approximation x_0 is close to α.

● Algebraic fractions

● To simplify algebraic fractions, first find the lowest common denominator,

e.g. for $\dfrac{1}{x} + \dfrac{2}{(1 - x)}$ the **lowest common**

denominator is $x(1 - x)$

so $\dfrac{1}{x} + \dfrac{2}{1 - x} \equiv \dfrac{(1 - x)}{x(1 - x)} + \dfrac{2x}{x(1 - x)} \equiv \dfrac{1 + x}{x(1 - x)}$

● The remainder theorem

If $a(x)$ and $b(x)$ are **polynomials** then $b(x)$ divides $a(x)$ provided the degree of $a(x)$ is greater than or equal to the degree of $b(x)$. The degree of the **remainder** is less than the degree of $b(x)$.

$$\dfrac{a(x)}{b(x)} = q(x) + \dfrac{r(x)}{b(x)}$$

remainder

quotient

If $r(x) = 0$ then $b(x)$ is a factor of $a(x)$.

Formulae you must know

● $p \Rightarrow q$ means p implies q (if p is true then so is q)

● $p \Leftarrow q$ means p is implied by q (if q is true then so is p)

● $p \Leftrightarrow q$ means p implies and is implied by q (p is equivalent to q)

Don't make these mistakes...

Be careful not to miss any terms when multiplying out brackets.

Be careful with the signs especially when there is a negative sign outside the bracket, e.g. $-(a - b) = -a + b$ **not** $-a - b$.

In a proof, remember that a result that is true for one particular value of a parameter is not necessarily true for all parameters, i.e. **proving a special case does not prove the general case**.

Q1 Express

$$\frac{5(x-3)(x+1)}{(x-12)(x+3)} - \frac{3(x+1)}{x-12}$$

as a fraction in its simplest form.

$$\frac{5(x-3)(x+1)}{(x-12)(x+3)} - \frac{3(x+1)(x+3)}{(x-12)(x+3)}$$

$$= \frac{(5x^2 - 10x - 15) - (3x^2 + 12x + 9)}{(x-12)(x+3)}$$

$$= \frac{2x^2 - 22x - 24}{(x-12)(x+3)}$$

$$= \frac{2(x-12)(x+1)}{(x-12)(x+3)}$$

$$= \frac{2(x+1)}{x+3}$$

- You need to write each part of the expression in terms of a common denominator.

- The common denominator that you need is $(x-12)(x+3)$.

- Now expand the brackets in the numerator. Leave the denominator as it is, in case you find you can cancel later.

- Remember to multiply each term in the first expression by 5 and each term in the second expression by 3.

- Simplify the numerator, then look for factors that might cancel with terms in the denominator.

Q2 The cubic polynomial $3x^3 - 7x^2 - 18x - 8$ is denoted by $f(x)$. Use the factor theorem to show that $(x+1)$ is a factor of $f(x)$.

Hence factorise $f(x)$ completely.

$$f(x) = 3x^3 - 7x^2 - 18x - 8$$

$$f(-1) = -3 - 7 + 18 - 8 = 0$$

$\therefore$ $(x+1)$ is a factor.

$$
\require{enclose}
\begin{array}{r}
3x^2 - 10x - 8 \\
x+1\enclose{longdiv}{3x^3 - 7x^2 - 18x - 8} \\
\underline{3x^3 + 3x^2} \\
-10x^2 - 18x - 8 \\
\underline{-10x^2 - 10x} \\
-8x - 8 \\
\underline{-8x - 8} \\
0
\end{array}
$$

$$f(x) = (x+1)(3x^2 - 10x - 8)$$
$$= (x+1)(3x+2)(x-4)$$

- Remember the factor theorem, which says that if $(x+a)$ is a factor of $f(x)$, then $f(-a) = 0$.

- You can either use long division here or 'trial and error' methods to take out $(x+1)$ as a factor.

- Now you can factorise the quadratic in the usual way, using the factors of $3x^2$ and -8.

Q3 **(a)** By sketching the curves with equations $y = 4 - x^2$ and $y = e^x$, show that the equation $x^2 + e^x - 4 = 0$ has one negative root and one positive root.

(b) Use the iteration formula $x_{n+1} = -\sqrt{4 - e^{x_n}}$ with $x_0 = -2$ to find in turn x_1, x_2 and x_3 and hence write down an approximation to the negative root of the equation, giving your answer to three decimal places.

An attempt to evaluate the positive root of the equation is made using the iteration formula

$x_{n+1} = \sqrt{4 - e^{x_n}}$ with $x_0 = 1.3$.

(c) Describe the result of such an attempt.

(a)

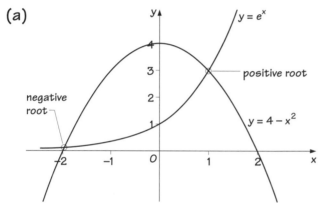

At intersection points $4 - x^2 = e^x$

$\Rightarrow 4 - x^2 - e^x = 0$ So there is one negative and one positive root.

(b) $x_{n+1} = -\sqrt{4 - e^{x_n}}$

$x_0 = -2$

$x_1 = -1.965\,875$

$x_2 = -1.964\,680$

$x_3 = -1.964\,637$

$x = -1.965$ is the negative root to 3 d.p.

- You can use your graphics calculator to produce the graphs, then copy them on to your solution sheet. Label each graph.

- Remember to answer the question you have been asked!

- Remember, to find an answer correct to three decimal places you need to work to more than three places of decimals. Enter -2 on your calculator, then use the ANS key to set up the iteration by keying $-\sqrt{(4 - \exp(\text{ANS}))}$.

(c) $x_0 = 1.3$

$x_1 = 0.57...$

$x_2 = 1.49...$

This leads to an error because $4 - e^{x^2}$ < 0 and x_3 involves the square root of a negative number. The iteration scheme fails.

- Your calculator may show an 'error' message. Think of a reason why this occurs.

Q4 Prove the following result.

'A triangle with sides that can be written in the form $n^2 + 1$, $n^2 - 1$, $2n$ (where $n > 1$) is right-angled.'

Show, by means of a counter example, that the converse is false.

For the triangle to be right-angled, it is necessary to show that:

$$(n^2 - 1)^2 + (2n)^2 = (n^2 + 1)^2$$

which is a statement of Pythagoras' theorem.

$\text{LHS} = (n^2 - 1)^2 + (2n)^2$

$\quad = n^4 - 2n^2 + 1 + 4n^2$

$\quad = n^4 + 2n^2 + 1$

$\quad = (n^2 + 1)^2$

$\quad = \text{RHS as required.}$

For example, a triangle with sides of 5, 12 and 13 units is right-angled.

But $12 = 2n \Rightarrow n = 6$

then $n^2 + 1 = 37$

and $n^2 - 1 = 35$

So this right-angled triangle does not have sides in the form $2n$, $n^2 + 1$, $n^2 - 1$.

- For problems involving proofs, you should state at the beginning the method you intend to use.

- In this problem you need to use Pythagoras' theorem.

- The converse statement says that for any right-angled triangle, the sides can be written as $(n^2 + 1)$, $(n^2 - 1)$, $2n$. To disprove the converse you simply need to find a Pythagorean triple which is not in this form.

- In problems like these, you should give a clear mathematical argument to support your answer.

- You just need to find a right-angled triangle that does not fit the rule.

Q1 The polynomial $f(x) = x^4 - x^2 - 2x + 2$ is denoted by $f(x)$. Show that $(x - 1)^2$ is a factor of $f(x)$. Hence factorise $f(x)$ completely.

Q2 Show that $(x - 1)$ is a factor of the cubic polynomial $f(x) = x^3 - 7x + 6$. Hence factorise $f(x)$ into the product of three linear factors.

Q3 $f(x) = 2x^3 + 5x^2 - 8x - 15$

    (a) Show that $(x + 3)$ is a factor of $f(x)$.

    (b) Hence factorise $f(x)$ as the product of a linear factor and a quadratic factor.

    (c) Find, to two decimal places, the two other values of x for which $f(x) = 0$.

Q4 What is the remainder when $x^4 + 2x^3 - x^2 + x - 4$ is divided by $(x^2 + 2)$?

Q5 Express $\dfrac{3x + 1}{x^2 + x + 1} + \dfrac{2}{x - 1} - \dfrac{3}{x + 2}$ as a single fraction in its simplest form.

Q6 An iterative sequence is defined by: $x_{n+1} = \dfrac{2x_n^3 + a}{3x_n^2}$ $(n = 0, 1, 2, ...)$ where $a > 0$, $x_0 = 1$.

    (a) Assuming that x_n tends to a limit L as $n \to \infty$ show that $L = \sqrt[3]{a}$.

    (b) Use this result to find $\sqrt[3]{3}$ correct to three decimal places, showing your working carefully.

Q7 In the expression $|x + 3| < 2|x| * x > 3$, $*$ is one of the symbols $\Rightarrow$, $\Leftarrow$ or $\Leftrightarrow$. State which of the symbols must be used and justify your answer.

Answers can be found on pages 112–113.

Key points to remember

• Definitions

If $y = a^x$ then $x = \log_a y$

• Special cases

$a = 10$: common logarithms
$$y = 10^x \Leftrightarrow x = \log y$$

$a = e$: natural logarithms
$$y = e^x \Leftrightarrow x = \ln y$$

Formulae you must know

- $\log_a 1 = 0 \quad a^0 = 1$
- $\log_a a = 1$
- $\log(uv) = \log u + \log v$
- $\log\left(\dfrac{u}{v}\right) = \log u - \log v$
- $\log(u^r) = r\log u$
- $a^{\log_a u} = u$

Don't make these mistakes ...

$\log(u + v)$ is **not** equal to $\log u + \log v$

$a^{u + v}$ is **not** equal to $a^u + a^v$

• Graphs of exponentials

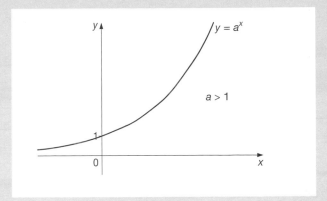

Exponential growth

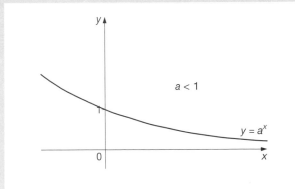

Exponential decay

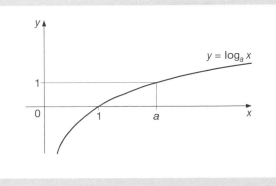

Q1 A function f is defined for the domain $x > 0$ by
$f(x) = \ln(x + 1)$.

(a) Find the range of f.

(b) Find an expression for $f^{-1}(x)$ in terms of x and e.

(a) $x = 0 \Rightarrow f(0) = \ln 1 = 0$

$x > 0 \Rightarrow f(x) > 1$

(b) $y = \ln(x + 1) \Rightarrow x + 1 = e^y$

$x = e^y - 1$

$f^{-1}(x) = e^x - 1$

- Start with $x = 0$ and simplify $\ln 1 = 0$.
- Remember that the domain of f is $x > 0$ and the log function is an increasing function.

- Remember to write the inverse function as $f^{-1}(x)$, i.e. in terms of x.

Q2 A function f is given by $f: x \rightarrow \ln(4 - 2x)$
$x \in \mathbb{R}, x < 2$.

(a) Find an expression for $f^{-1}(x)$.

(b) Sketch the curve with equation $y = f^{-1}(x)$, showing the coordinates of the points where the curve meets the axes.

The function g is given by $g: x \rightarrow 3^x \quad x \in \mathbb{R}$.

(c) Find the value of x for which $g(x) = 1.5$, giving your answer to three decimal places.

(d) Evaluate $g(f(1))$.

(a) $y = \ln(4 - 2x) \quad x < 2$

$4 - 2x = e^y$

$x = 2 - \frac{1}{2}e^y$

$f^{-1}(x) = 2 - \frac{1}{2}e^x$

- Remember to write the inverse function in terms of x.

(b) $x = 0 \rightarrow f^{-1}(0) = 2 - \frac{1}{2}e^0 = 2 - \frac{1}{2} = \frac{3}{2}$

$y = 0 \rightarrow 2 - \frac{1}{2}e^x = 0 \rightarrow e^x = 4 \rightarrow x = \ln 4$

- First, you need to find the coordinates of the points of intersection with the x- and y-axes. Remember to show these coordinates on your graph.

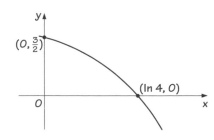

Graph of $y = f^{-1}(x)$

(c) $g(x) = 1.5 \Rightarrow 3^x = 1.5$

Taking logs of each side:

$\log 3^x = \log 1.5 \Rightarrow x \log 3 = \log 1.5$

$\Rightarrow x = \dfrac{\log 1.5}{\log 3} = 0.369$ (to 3 d.p.)

- Remember to use logs to base 10 or base e because they are on your calculator.

Q3 **(d)** $g(f(1)) = g(\ln 2) = 3^{\ln 2} = 2.141$

- Evaluate f(1) first, then g(f(1)).

At t minutes after an oven is switched on, its temperature $\theta\,°C$ is given by:

$$\theta = 200 - 180e^{-0.1t}$$

(a) State the value approached by the temperature in the oven after a long time.

(b) Find the time taken for the oven's temperature to reach 150 °C.

(c) Find the rate at which the temperature is increasing at the instant when it reaches 150 °C.

(a) As $t \to \infty$ $e^{-0.1t} \to 0$

$\Rightarrow \theta \to 200\,°C$

- You should remember how exponential functions behave.

(b) When $\theta = 150\,°C$:

$200 - 180e^{-0.1t} = 150$

$180e^{-0.1t} = 50$

$e^{-0.1t} = \dfrac{50}{180} = \dfrac{5}{18}$

$-0.1t = \ln \dfrac{5}{18}$

$t = 12.8$ minutes

- Take care with your signs. Remember that $\ln x < 0$ for $x < 1$ so the negative signs cancel.

(c) The rate of increase of temperature is $\dfrac{d\theta}{dt}$.

$\dfrac{d\theta}{dt} = 18e^{-0.1t}$

$\theta = 150°$ at $t = 12.8$ minutes

$\dfrac{d\theta}{dt} = 18e^{-0.1 \times 12.8} = 5\,°C$ per minute

- Now you can write down $e^{-0.1t} = \frac{5}{18}$ directly from part (b).
- So $18e^{-0.1t} = 18 \times \frac{5}{18} = 5$

Q1 Given that $f(x) = 4^x$ show that each of these relations is true.

(a) $f(0) = 1$ (b) $f(x + y) = f(x)f(y)$ (c) $\dfrac{f(x)}{f(y)} = f(x - y)$ (d) $\{f(x)\}^n = f(nx)$

Q2 Simplify each of these.

(a) $\frac{1}{3}\ln e^3$ (b) $6\ln \sqrt[3]{e}$ (c) $7\ln e^4 - 6\ln\sqrt{e}$ (d) $e^{4\ln x}$ (e) $\ln e^{2x}$

Q3 The population, P, of a certain city at time t (measured in years) is given by the formula:

$P = 50\,000e^{0.05t}$

Find the time taken for the population to:

(a) double
(b) increase to 55 000.

Q4 Solve the equation $e^{4x} + e^{2x} - 6 = 0$.

Q5 Solve these equations.

(a) $\ln x = 7$ (b) $3 - 4\ln x = 0$ (c) $3\ln x^3 + 4\ln x = 9$

(d) $\ln 7x + \ln 5x = 1$ (e) $\ln(2x - 3) - \ln x = 0$

Q6 Bacteria are growing rapidly according to the law $N = N_0 e^{\kappa t}$ where N is the number of bacteria present at time t hours and N_0 and κ are constants.

If the number of bacteria grows from 200 000 to 300 000 in five hours, how many bacteria do you predict after ten hours?

Q7 The function f is defined as $f(x) = e^x - 5x$.

(a) Determine $f'(x)$.

(b) Find the value of x for which $f'(x) = 0$, giving your answer to two decimal places.

(c) Show, by calculation, that there is a root, α, of the equation $f(x) = 0$ such that $0.2 < \alpha < 0.3$.

(d) Determine the integer P such that the other root, β, of the equation $f(x) = 0$ lies in the interval: $\frac{1}{10}P < \beta < P + \frac{1}{10}$.

Answers can be found on pages 113–114.

Key points to remember

- A **tangent to a curve** has the equation $y = mx + c$, where m is found by evaluating $\dfrac{dy}{dx}$ at the point where the tangent touches the curve.

- A **normal to a curve** has the equation $y = mx + c$, where m is found by evaluating $-1/\dfrac{dy}{dx}$ at the point that the normal crosses the curve.

- The **second derivative** is found by **differentiating** $\dfrac{dy}{dx}$ and is written as $\dfrac{d^2y}{dx^2}$.

- The **rate of change** of a quantity x, with respect to t, is given by $\dfrac{dx}{dt}$.

- **Stationary points**

  At a stationary point $\dfrac{dy}{dx} = 0$.

  If $\dfrac{d^2y}{dx^2} < 0$ the stationary point will be a local maximum.

  If $\dfrac{d^2y}{dx^2} > 0$ the stationary point will be a local minimum.

  If $\dfrac{d^2y}{dx^2} = 0$ the nature of the stationary point must be determined by examining the gradient of the curve either side of the stationary point.

Formulae you must know

- $\dfrac{d}{dx}(e^{kx}) = ke^{kx}$

- $\dfrac{d}{dx}(\ln(kx)) = \dfrac{1}{x}$

Don't make these mistakes...

Don't assume that **if $\dfrac{d^2y}{dx^2} = 0$ at a stationary point**, then it must be a point of inflexion. (It could be a local minimum, a local maximum or a point of inflexion.)

Don't confuse **differentiating and integrating** e^{kx}. (Multiply by k when differentiating.)

Don't forget to **check the nature** of a stationary point.

Q1 Show that $y = e^{4x} - 8x$ has a local minimum at $x = \frac{1}{4}\ln 2$.

$y = e^{4x} - 8x$

So $\dfrac{dy}{dx} = 4e^{4x} - 8$

At stationary points $\dfrac{dy}{dx} = 0$

So $4e^{4x} - 8 = 0$

$4e^{4x} = 8$

$e^{4x} = 2$

$4x = \ln 2$

$x = \frac{1}{4}\ln 2$

In this case $\dfrac{d^2y}{dx^2} = 16e^{4x}$.

As this will be greater than 0 for all x, the value found above must give a local minimum.

- At a **stationary point** $\dfrac{dy}{dx} = 0$, so the first step is to **find** $\dfrac{dy}{dx}$ and solve the equation $\dfrac{dy}{dx} = 0$.

- This value of x may or may not give a ~~local minimum~~. To identify the type of stationary point you need to **consider** $\dfrac{d^2y}{dx^2}$.

- Remember that $\dfrac{d^2y}{dx^2} > 0$ for a local **minimum**.

Q2 If $f(x) = e^{2x} - 5x$ find the range of values of x for which $f(x)$ is an increasing function.

$f(x) = e^{2x} - 5x$

So $f'(x) = 2e^{2x} - 5$

For an increasing function:

$2e^{2x} - 5 > 0$

$e^{2x} > \frac{5}{2}$

$x > \frac{1}{2}\ln\frac{5}{2}$

- For an **increasing function** you need $f'(x) > 0$, so first **differentiate to find** $f'(x)$.

- Then **form and solve the inequality** $f'(x) > 0$.

- You should **work in exact form** when possible.

KATE WONDERED WHETHER THIS WAS SIDNEY'S NORMAL WAY OF SURFING

Q3 (a) Find the equation of the tangent to the curve $y = \ln 2x + 5x + 3$, at the point where $x = \frac{1}{2}$.

(b) Find the equation of the normal to the curve at the same point.

(a) When $x = \frac{1}{2}$, $y = \ln 1 + \frac{5}{2} + 3 = \frac{11}{2}$.

$y = \ln 2x + 5x + 3$

So $\dfrac{dy}{dx} = \dfrac{1}{x} + 5$

When $x = \frac{1}{2}$, $\dfrac{dy}{dx} = \dfrac{1}{\frac{1}{2}} + 5 = 7$

So the gradient of the tangent is 7 and its equation will be $y = 7x + c$.

Using $x = \frac{1}{2}$ and $y = \frac{11}{2}$, gives:

$\dfrac{11}{2} = 7 \times \dfrac{1}{2} + c \quad \Rightarrow \quad c = 2$

The equation of the tangent is $y = 7x + 2$.

(b) The equation of the normal will be $y = -\frac{1}{7}x + c$.

Using $x = \frac{1}{2}$ and $y = \frac{11}{2}$, gives:

$\dfrac{11}{2} = -\dfrac{1}{7} \times \dfrac{1}{2} + c \quad \Rightarrow \quad c = \dfrac{39}{7}$

The equation of the normal is $y = -\frac{1}{7}x + \frac{39}{7}$.

- **First you need to find the coordinates** of the point at which the **tangent meets the curve**.

- The **gradient** of the curve at $x = \frac{1}{2}$ is given by the **derivative**, so you need to differentiate and substitute $x = \frac{1}{2}$.

- Use the **equation of a straight line** $y = mx + c$.

- Use the **coordinates** to calculate the **value of the constant** c.

- The **gradient of the normal** is given by $-1 \big/ \dfrac{dy}{dx}$.

- Use the **coordinates** to calculate the value of the constant c.

Q4 The level of radioactivity of a substance is given by $R = 40e^{-0.3t}$. Find the rate at which the level of radioactivity is decreasing when:

(a) $t = 20$ (b) $R = 20$.

$R = 40e^{-0.3t}$

So $\dfrac{dR}{dt} = -0.3 \times 40e^{-0.3t} = -12e^{-0.3t}$

(a) When $t = 20$, $\dfrac{dR}{dt} = -12e^{-6} = -0.0297$

correct to three significant figures.

(b) When $R = 20$, $20 = 40e^{-0.3t}$

$e^{-0.3t} = \dfrac{1}{2}$

$t = -\dfrac{10}{3} \ln \dfrac{1}{2} = \dfrac{10}{3} \ln 2$

When $t = \dfrac{10}{3}\ln 2$, $\dfrac{dR}{dt} = -12e^{-\ln 2} = -6$.

- **The rate of change** will be given by $\dfrac{dR}{dt}$, so find this first.

- **Substitute** $t = 20$, to find the **required rate of change**.

- In the second case you need to find the value of t when $R = 20$. To do this, **substitute** $R = 20$ and **solve for** t.

- Then substitute the value of t into $\dfrac{dR}{dt}$.

- Alternatively, you could use $\dfrac{dR}{dt} = -0.3R = -0.3 \times 20 = -6$.

Questions to try

Q1 A curve has equation $y = x^4 + x^6$.

    (a) Find the values of $\dfrac{dy}{dx}$ and $\dfrac{d^2y}{dx^2}$ when $x = 0$.

    (b) Determine the nature of the stationary point at $x = 0$.

Q2 Show that $y = ax - \ln x$, where a is a positive constant, has a local minimum at $x = \dfrac{1}{a}$.

Q3 If $f(x) = x + 2e^{-x}$, find the range of values of x for which $f(x)$ is decreasing.

Q4 The equation of a curve is $y = \dfrac{1}{x} + \ln 2x$, $x > 0$.

    (a) Find $\dfrac{dy}{dx}$ and $\dfrac{d^2y}{dx^2}$.

    (b) Calculate the gradient of the curve when $x = 2$.

    (c) Find the coordinates of the stationary point on the curve and determine the nature of this stationary point.

Q5 A curve has equation $y = 2x + 3e^{-2x}$.

    (a) Find the coordinates of the stationary point on this curve.

    (b) Find $\dfrac{d^2y}{dx^2}$.

    (c) Determine the nature of the stationary point on the curve.

Q6 **(a)** Given that $y = \ln 4x - x^2$, $x > 0$, find $\dfrac{d^2y}{dx^2}$.

    (b) Find the exact coordinates of the stationary point on the curve $y = \ln 4x - x^2$ and determine the nature of the stationary point.

Q7 The equation of a curve is $y = x + \ln 2x$, $x > 0$.
Find the equation of the tangent to this curve at the point $x = 1$.

Q8 The curve with equation $y = 2 - e^{2x}$ intersects the y-axis at the point A.

    (a) Find the equation of the tangent to the curve at A.

    (b) Find the coordinates of the point where the normal to the curve at A intersects the x-axis.

Q9 Find the equation of the normal to the curve with equation $y = 5 - 2\ln x$, at the point where $x = e$.

Q10 The number of cells, N, produced during time t in an experiment is given by $N = 120e^{0.05t}$.

    (a) Find the rate at which N is increasing, when $t = 20$.

    (b) Find the rate at which N is increasing, when $N = 360$.

Answers can be found on pages 114–116.

11 Integration II

Key points to remember

- Integrate **exponential functions** using the formula
  $$\int e^{kx}dx = \frac{1}{k}e^{kx} + c.$$

- Integrate to find **areas under curves**.

- **Volumes of revolution** are produced when a region is rotated through 360° around an axis, normally the *x*-axis.

For example this graph shows the region enclosed by the curve $y = \sqrt{x}$, the line $x = 4$ and the *x*-axis.

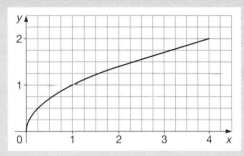

When this some region is rotated through 360° around the *x*-axis the solid illustrated on the right is formed.

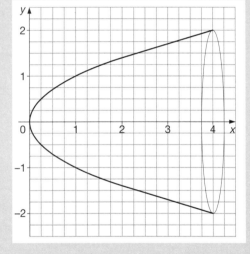

The volume of this solid is given by

$$\int_0^4 \pi(\sqrt{x})^2 dx = \int_0^4 \pi x\, dx = \left[\frac{\pi x^2}{2}\right]_0^4 = 8\pi$$

Don't make these mistakes ...

Don't forget to **square y when finding volumes of revolution**.

Don't forget to **include a constant of integration**.

Don't forget to **include π when finding a volume of revolution**.

Don't confuse the **formulae for integrating and differentiating e^{kx}**.

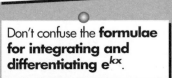

Q1 Find the area of the finite region bounded by the curve $y = 2e^x$, the line $x = 4$ and the axes, giving your answer in exact form.

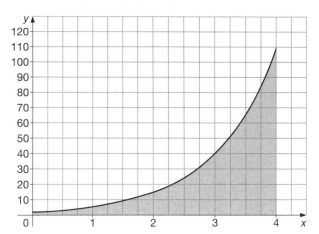

Area $= \displaystyle\int_0^4 2e^x dx$

$\qquad = \left[2e^x \right]_0^4$

$\qquad = 2e^4 - 2e^0$

$\qquad = 2e^4 - 2$

- Make sure that you know what the **limits of integration** are, using **a sketch** to help if necessary.

- Remember that $\displaystyle\int e^x dx = e^x + c$.

- Use the **square bracket notation after integrating**, and **no constant of integration**.

- Whenever you can, give your answer in **exact form**, in this case in terms of e.

- Remember that $e^0 = 1$.

Q2 The finite region bounded by the curve $y = 5x - x^2$ and the x-axis is rotated through $360°$ around the x-axis. Find the volume of the solid that is formed.

Solving $5x - x^2 = 0$, gives:
$5x - x^2 = 0$
$x(5 - x) = 0$
$x = 0$ or $x = 5$

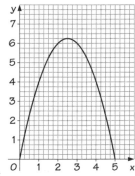

Volume $= \displaystyle\int_0^5 \pi(5x - x^2)^2 dx$

$\qquad = \pi \displaystyle\int_0^5 (25x^2 - 10x^3 + x^4) dx$

$\qquad = \pi \left[\dfrac{25x^3}{3} - \dfrac{10x^4}{4} + \dfrac{x^5}{5} \right]_0^5$

$\qquad = \pi \left(\dfrac{3125}{3} - \dfrac{3125}{2} + 625 \right)$

$\qquad = \dfrac{625\pi}{6}$

- First you need to **find the points where the curve intersects the x-axis**, by solving the equation $y = 0$.

- The two values of x give the **limits of integration**.

- The diagram shows the region that is to be rotated.

- Use the formula $\displaystyle\int_0^5 \pi y^2 dx$ to find the **volume of the solid**.

- **Multiply out the brackets before attempting to integrate**.

- Remember to **integrate each term of the expression**.

- You can **substitute the limits of integration** to find the actual volume.

- The value here has been found as a fraction and a multiple of π. **It is good policy to work in fractions**, like this, whenever possible.

Q3 A curve passes through the point P with coordinates $(\ln 6, 1)$. The gradient of the curve is given by $e^{-2x} + 3$. Find the equation of the curve.

Integrating gives:

$$y = \int (e^{-2x} + 3)dx$$
$$= -\frac{1}{2}e^{-2x} + 3x + c$$

At P $x = \ln 6$ and $y = 1$, so:

$$1 = -\frac{1}{2}e^{-2\ln 6} + 3\ln 6 + c$$
$$1 = -\frac{1}{2}e^{\ln(\frac{1}{36})} + 3\ln 6 + c$$
$$1 = -\frac{1}{72} + 3\ln 6 + c$$
$$c = \frac{73}{72} - 3\ln 6$$

So $y = -\frac{1}{2}e^{-2x} + 3x + \frac{73}{72} - 3\ln 6$

Q4 If the area enclosed by the curve $y = e^{2x}$, the x-axis, the y-axis and the line $x = a$ is 2 find the value of a.

This area is given by:

$$\int_0^a e^{2x}dx = \left[\frac{1}{2}e^{2x}\right]_0^a$$
$$= \frac{1}{2}(e^{2a} - 1)$$
$$2 = \frac{1}{2}(e^{2a} - 1)$$
$$4 = e^{2a} - 1$$
$$e^{2a} = 5$$
$$a = \frac{1}{2}\ln 5$$

How to score full marks

- Integrating the expression for the gradient will give the **equation of the curve**.
- Don't forget to include the constant c.
- **Substitute** the values of x and y at the point P, to calculate the value of c.
- You can use the **rules of logarithms** to simplify the equation.
- The result here has been given in exact form.

- This question gives the area and asks you to find the **upper limit of the integral**, which is a in this case.
- The diagram shows the area that is equal to 2.
- Begin by **setting up the integral** with the **lower limit** as 0 and the **upper limit** as a.
- As the area is 2, you can form **an equation using the result of the integration**.
- As the equation contains an **exponential term**, you must use **logarithms** to solve the equation.

51

Questions to try

Q1 Find the area of the finite region bounded by the axes, the curve $y = e^x - 2x$ and the line $x = 3$, giving your answer in terms of e.

Q2 Find the area of the region bounded by the curve $y = e^{-x}$ and the lines $y = x + 1$, $x = 1$ and the x-axis.

Q3 Find the area of the region bounded by the curve $y = e^{3x} + 1$, the lines $x = 0$, $x = 2$ and the x-axis.

Q4 A curve intersects the x-axis at the point with coordinates $(\frac{1}{2}(\ln 3 - \ln 2), 0)$. The gradient of the curve is given by $2e^{\frac{1}{2}x}$. Determine the equation of the curve.

Q5 The region enclosed the lines $x = 1$, $x = 4$, the curve $y = \dfrac{1}{x}$ and the x-axis is rotated through $360°$ around the x-axis. Find the volume of the solid that is created, in terms of π.

Q6 The region enclosed by the lines $x = 1$, $x = 4$, the curve $y = \sqrt{x^2 + 1}$ and the x-axis is rotated through $360°$ around the x-axis. Find the volume of the solid that is created.

Q7 The shaded region, A, shown in the diagram is bounded by the curve $y = \sqrt{x - 1}$, the line $x = 3$ and the x-axis. Find the volume formed when A is rotated through $360°$ about the x-axis.

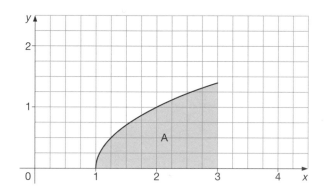

Q8 The finite region bounded by the curve with equation $y = x - x^2$ and the x-axis is rotated through $360°$ about the x-axis. Using integration, find, in terms of π, the volume of its solid form.

Q9 Find, in terms of e, the area of the finite region bounded by the curve with equation $y = \frac{1}{2}e^x$, the coordinate axes and the line with equation $x + 2 = 0$.

Q10 A curve has equation $y = f(x)$. The curve passes through the point with coordinates $(\ln 2, 5(1 + \ln 2))$ and $\dfrac{dy}{dx} = 2e^{2x} + 5$. Find $f(x)$, in exact form.

Q11 The region bounded by the axes, the line $x = k$, where k is a positive constant and the curve $y = e^{-3x}$ has area $\dfrac{7}{24}$. Find k in exact form.

Answers can be found on pages 116–117.

Key points to remember

● **Transformations**

1

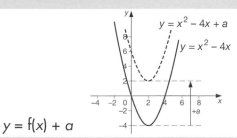

$y = f(x) + a$

A translation of $y = f(x)$ through a units, parallel to the y-axis

2

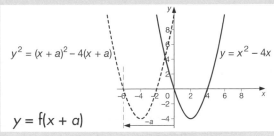

$y = f(x + a)$

A translation of $y = f(x)$ through $-a$ units, parallel to the x-axis

3

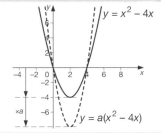

$y = af(x)$

A stretch of $y = f(x)$, scale factor a, parallel to the y-axis

4

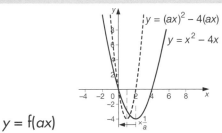

$y = f(ax)$

A stretch of $y = f(x)$, scale factor $\frac{1}{a}$, parallel to the x-axis

The special cases when $a = -1$ in 3 and 4 lead to two more important geometrical transformations:

5

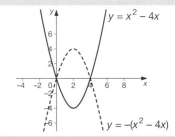

$y = -f(x)$

A reflection of $y = f(x)$ in the x-axis

6

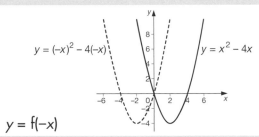

$y = f(-x)$

A reflection of $y = f(x)$ in the y-axis

● An **even function** $f(x)$ has the property $f(-x) = f(x)$.

● An **odd function** $f(x)$ has the property $f(-x) = -f(x)$.

Formulae you must know

● $f(x + a)$ translates $f(x)$ to the left by a units

● $f(x - a)$ translates $f(x)$ to the right by a units

● $f(x) + a$ translates $f(x)$ up by a units.

Don't make these mistakes ...

Don't translate the graph the wrong way.

Q1 The function f is defined for all real values of x by $f(x) = 2(x-1)^2 + 4$.

Describe geometrically a series of transformations whereby the graph of $y = f(x)$ can be obtained from the graph of $y = x^2$.

Translate the curve one unit to the right to form $y = (x-1)^2$.

Stretch, scale factor 2, parallel to the y-axis to form $y = 2(x-1)^2$.

Translate four units upwards, parallel to the y-axis to form $y = 2(x-1)^2 + 4$.

- First, you need to split the transformation into three parts.

- Remember that changing x to $(x-1)$ shifts the curve to the right.

Q2 The graph of $y = f(x)$ is sketched below, where a is a positive constant.

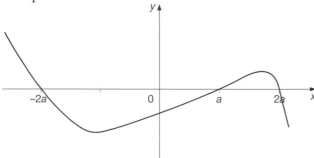

On separate sets of axes sketch the graphs of the following equations, indicating clearly the intercepts with the x-axis.

(a) $y = f(x + a)$ **(b)** $y = f(-x)$
(c) $y = f(2x)$ **(d)** $y = |f(x)|$

(a)

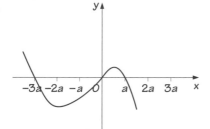

- You obtain full marks for showing the correct shape and the intercepts on the x-axis.

(b)

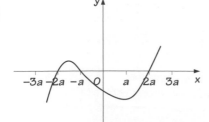

- $f(-x)$ is a reflection in the y-axis.

(c)

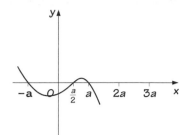

- f(2x) is a stretch, scale factor $\frac{1}{2}$, parallel to the x-axis.

(d)

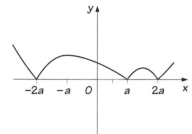

- For |f(x)| all negative parts of f(x) are reflected in the x-axis.

Q3 Sketch the graph of $y = \cos x$ for values of x from 0 to 360°.

Sketch on the same diagram, the graph of $y = \cos(x - 60°)$.

Use your diagram to solve the equation $\cos x = \cos(x - 60°)$ for values of x between 0 and 360°. Indicate clearly on your diagram how the solutions relate to the graphs.

State how many values of x satisfying the equation $\cos 10x = \cos(10x - 60°)$ lie between 0 and 360°.

- You should explain your reasoning briefly, but no further detailed working or sketching is necessary.

- First you need to draw cos x correctly and remember that cos(x − 60°) is a translation of 60° to the right.

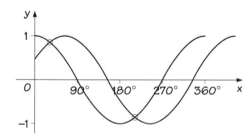

x = 30° and x = 210°

For cos10x and cos(10x − 60°) the graphs are 'squashed' so that 10 complete cycles fit the interval 0° to 360°. Hence there are 20 solutions.

- Show the points of intersection of the two graphs clearly.

- Remember the symmetry of cos x; cos(−30°) = cos 30°.

Q1 The function f is defined for all real values of x as $f(x) = e^{4(x-2)}$. Describe geometrically a series of transformations whereby the graph of $y = f(x)$ can be obtained from the graph of $y = e^x$.

Q2 The figure shows a sketch of the curve with equation $y = f(x)$.

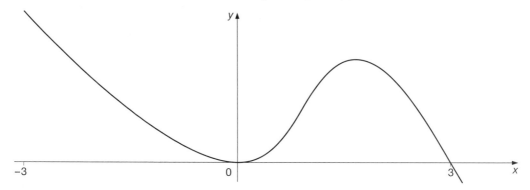

In separate diagrams show, for $-3 \leqslant x \leqslant 3$ a sketch of the curve with equation:

(a) $y = f(-x)$
(b) $y = -f(x)$
(c) $y = f(|x|)$

marking on each sketch the x-coordinates of any point, or points, where a curve touches or crosses the x-axis.

Q3 The function f has as its domain the set of all non-zero real numbers, and is given by $f(x) = \dfrac{1}{x}$ for all x in this set. On a single diagram, sketch each of the following graphs, and indicate the geometrical relationships between them.

(a) $y = f(x)$ **(b)** $y = f(x + 1)$ **(c)** $y = f(x + 1) + 2$

Deduce, explaining your reasoning, the geometrical relationship between the graph of $y = \dfrac{2x + 3}{x + 1}$ and the graph of $y = \dfrac{1}{x}$.

Q4 The function $f(x)$ is defined for all values of x except $x = 0$ and is an odd function, i.e. $f(-x) = -f(x)$.

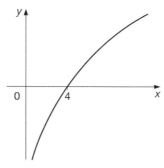

(a) Part of the graph of $y = f(x)$ is shown on the diagram. Copy and complete the sketch.

(b) Draw a separate sketch to illustrate the graph of $y = f(x + 3)$, showing clearly where the graph will intercept the x-axis.

Answers can be found on pages 117–118.

Key points to remember

Constant acceleration formulae

- You will need to know the four formulae and remember that they can only be applied when the acceleration is constant.

Graphs

- The **displacement** of a particle is given by the area enclosed by a velocity-time graph.

- The **acceleration** is given by the gradient of a velocity-time graph.

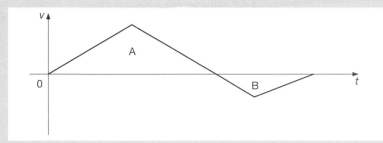

- On this graph:

  the area of triangle A gives the displacement in the **positive direction**

  the area of triangle B gives the displacement in the **negative direction**

  total displacement = area of A − area of B

- The **gradient** of a displacement-time graph gives the **velocity**.

Formulae you must know

- $v = u + at$
- $s = ut + \frac{1}{2}at^2$
- $v^2 = u^2 + 2as$
- $s = \frac{1}{2}(u + v)t$

Don't make these mistakes...

Don't use constant acceleration equations when the **acceleration is not constant**.

Don't forget to consider the **direction of motion**, when finding displacements from a velocity-time graph.

Q1 A lift rises from rest, accelerating at 0.2 m s^{-2} for 2 seconds, then travels at a constant speed for 5 seconds and slows down over a 3-second period.

Find the total distance travelled by the lift.

The speed reached by the lift after 2 seconds is:

$v = 0 + 0.2 \times 2$

$\quad = 0.4\ \text{m s}^{-1}$

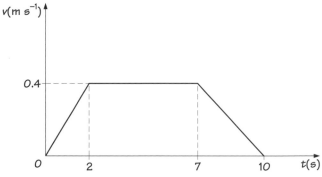

From the velocity-time graph:

displacement $= \frac{1}{2}(5 + 10) \times 0.4$

$\quad\quad\quad\quad\quad = 3\ \text{m}$

- First, you need to calculate the speed at the end of the first 2 seconds using:

$$v = u + at$$

- It is a good idea to draw a sketch graph in questions like this. Then you can find the distance by calculating the area under (or enclosed by) the graph.

- You can find the area by using the formula for the area of a trapezium.

$$A = \frac{1}{2}(a + b)h$$

- Alternatively you could split the area into a rectangle and two triangles.

Q2 A ball is thrown vertically upwards, from a height of 1.2 m, with an initial velocity of 4 m s^{-1}. Assume that no resistance forces act on the ball.
(a) Find the maximum height of the ball above the ground.
(b) Find the speed of the ball when it hits the ground.

(a) Assume the upward direction is positive. At its maximum height $v = 0$ so:

$0^2 = 4^2 + 2 \times (-9.8)s$

$s = \dfrac{4^2}{2 \times 9.8} = 0.82\ \text{m}$

The maximum height above the ground

$= 0.82 + 1.2$

$= 2.02\ \text{m}$

(b) The ball hits the ground when $s = -1.2$.

$v^2 = 4^2 + 2 \times (-9.8) \times (-1.2)$

$v = 6.29\ \text{m s}^{-1}$

- In questions like this, you should define the positive direction. In this case, the student has defined it as 'upwards'.

- You can take the acceleration (due to gravity) as $-9.8\ \text{m s}^{-2}$ and use the initial velocity, which is stated in the question as $4\ \text{m s}^{-1}$.

- Now use these values in the formula:

$$v^2 = u^2 + 2as$$

- Remember that the ball was not launched at ground level, so you must add on the extra height.

- When the ball hits the ground it will be 1.2 m below its release point, so s will be -1.2.

Q3 The graph below is a velocity-time graph for a train.

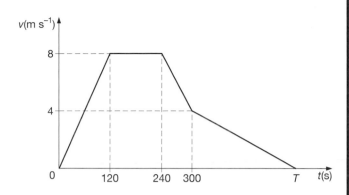

(a) Find the total distance travelled by the train in the first 300 seconds.

(b) Find T if the train travels a total distance of 2 km.

(a) Taking each stage of the motion:

for $t = 0$ to 120,
distance $= \frac{1}{2} \times 120 \times 8 = 480$ m

for $t = 120$ to 240,
distance $= 120 \times 8 = 960$ m

for $t = 240$ to 300,
distance $= \frac{1}{2} \times (8 + 4) \times 60 = 360$ m

Total distance $= 480 + 960 + 360$
$= 1800$ m

(b) The distance travelled in the final stage must be $2000 - 1800 = 200$ m.

$200 = \frac{1}{2} \times (T - 300) \times 4$

$200 = 2T - 600$

$T = 400$

- Remember that for a velocity-time graph, the distance covered is the area under the graph.

- The total area is made up of a triangle, a rectangle and a trapezium.

- You can use the formulae for each of these areas to find the total distance travelled.

- The distance travelled on the final stage is given by the area of the triangle. You can use the result from part (a) to find the actual distance.

- Then you can calculate the length of the base of the triangle as $T - 300$.

Questions to try

Q1 A ball is thrown upwards at a speed of $6\,\mathrm{m\,s}^{-1}$, from a height of $5\,\mathrm{m}$.

    (a) Find the maximum height of the ball.   **(b)** Find the time that the ball is in the air.
    (c) Find the speed of the ball when it hits the ground.

Q2 A car starts from rest at a point O and moves in a straight line. The car moves with a constant acceleration $4\,\mathrm{m\,s}^{-2}$ until it passes the point A, when it is moving with speed $10\,\mathrm{m\,s}^{-1}$. It then moves with constant acceleration $3\,\mathrm{m\,s}^{-2}$ for $6\,\mathrm{s}$ until it reaches the point B. Find:

    (a) the speed of the car at B   **(b)** the distance OB.

Q3 A train T_1 moves from rest at station A with constant acceleration $2\,\mathrm{m\,s}^{-2}$ until it reaches a speed of $36\,\mathrm{m\,s}^{-1}$. It maintains this constant speed for $90\,\mathrm{s}$ before the brakes are applied, which produces a constant retardation of $3\,\mathrm{m\,s}^{-2}$. The train T_1 comes to rest at station B.

    (a) Sketch a speed-time graph to illustrate the journey of T_1 from A to B.
    (b) Show that the distance between A and B is $3780\,\mathrm{m}$.

A second train T_2 takes $150\,\mathrm{s}$ to move from rest at A to rest at B. The diagram shows the speed-time graph illustrating this journey.

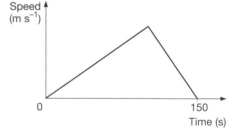

    (c) Explain briefly one way in which T_1's journey differs from T_2's journey.
    (d) Find the greatest speed, in $\mathrm{m\,s}^{-1}$, attained by T_2 during its journey.

Q4 The graph shows how the velocity of a toy train varies on one section of a track.

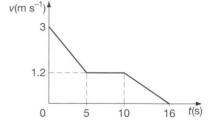

    (a) Describe briefly the movement of the train.
    (b) Find the total distance travelled by the train.
    (c) Find the distance travelled by the train when $t = 12$.

Q5 A lift travels from rest from the ground floor and comes to rest again at a car park 15 metres above the ground floor. The motion of the lift takes place in three stages. In the first stage the lift moves with a constant acceleration, it then moves with a constant velocity of $4\,\mathrm{m\,s}^{-1}$ and finally it moves with a constant retardation until it comes to rest. The times for the three stages of the motion are $1\frac{1}{2}$, t and 1 seconds, respectively.

    (a) Sketch a velocity-time graph to show the motion of the lift.
    (b) Hence, or otherwise, calculate the time for which the lift is in motion.
    (c) Calculate the average velocity of the lift during the motion.

Q6 A car, travelling at $20\,\mathrm{m\,s}^{-1}$, passes a point, A. The car moves from A with a constant acceleration of $2\,\mathrm{m\,s}^{-2}$ until it reaches the point B. It then moves from B to C at a constant speed in a time of 10 seconds. The car travels a total of $425\,\mathrm{m}$.

    (a) Find the time that the car takes to travel from A to B.
    (b) Find the speed of the car between B and C.

Answers can be found on pages 118–119.

Key points to remember

- When an object such as a boat moves in a current, the resultant velocity is the sum of its own velocity in still water and the velocity of the current.

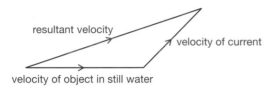

velocity of object in still water

resultant velocity

velocity of current

You can solve problems of this type by scale drawing, but it is generally better to use trigonometry.

- The constant acceleration equations can be used in two dimensions (directions at right angles to each other), where the velocity, acceleration and position are expressed as vectors.

- When an object moves with a constant velocity **v** from an initial position $\mathbf{r}_0$, then at time t its position will be given by $\mathbf{v}t + \mathbf{r}_0$.

- Velocities expressed with bearings can be written in terms of the unit vectors **i** and **j**. For example, a velocity of $40 \, \text{m s}^{-1}$ on bearing of $070°$ can be written as $40 \sin 70° \mathbf{i} + 40 \cos 70° \mathbf{j}$.

- Average velocity is the displacement divided by the time taken. This may not be the same as the actual velocity. For example, if an object completes a circle in 20 seconds, the displacement will be zero because the motion begins and ends at the same point. The average velocity will be zero, but clearly the actual velocity will not have been zero.

Formulae you must know

- $\mathbf{v} = \mathbf{u} + \mathbf{a}t$
- $\mathbf{r} = \mathbf{u}t + \frac{1}{2} \mathbf{a}t^2$
- $\mathbf{r} = \frac{1}{2} (\mathbf{u} + \mathbf{v})t$
- average velocity = $\dfrac{\text{displacement}}{\text{time}}$

Don't make these mistakes...

Don't forget to express quantities as **vectors**.

Don't forget to take account of **initial positions**.

Don't omit **i** or **j** from expressions.

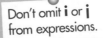

Q1 Two boats, A and B, set out from the same port at the same time. One travels at a speed of 8 m s^{-1} on a bearing of 330° and the other travels at 5 m s^{-1} due west. The port is taken to be the origin.

(a) Find expressions for the positions of each boat, relative to their starting points, using the unit vectors **i** and **j**, which are directed east and north respectively.

(b) Determine when the distance between the two boats is equal to 1.4 km.

(a) Velocity of A $= \underline{v}_A = -8\sin 30°\underline{i} + 8\cos 30°\underline{j}$

$$= -8 \times \tfrac{1}{2}\underline{i} + 8 \times \frac{\sqrt{3}}{2}\underline{j}$$

$$= -4\underline{i} + 4\sqrt{3}\underline{j}$$

$\underline{r}_A = -4t\underline{i} + 4\sqrt{3}t\underline{j}$

$\underline{v}_B = -5\underline{i}$

$\underline{r}_B = -5t\underline{i}$

- Start by expressing the velocities as vectors. You should leave the surds in the expressions, so that you don't introduce any rounding errors.
- The positions of the boats are given by $\mathbf{v}t + \mathbf{r_0}$ where $\mathbf{v}$ is the velocity and $\mathbf{r_0}$ is the initial position.

(b) Distance between the boats

$$= \sqrt{(-4t - (-5t))^2 + (4\sqrt{3}t)^2}$$

$$= \sqrt{t^2 + 48t^2} = \sqrt{49t^2}$$

$$= 7t$$

When $d = 1400$, $1400 = 7t$,

so $t = 200$ seconds.

- To find the distance between the two boats, you need to calculate the magnitude of the vector $\mathbf{r}_A - \mathbf{r}_B$.
- Remember that the velocities are given in m s^{-1}, so don't forget to change the distance of 1.4 km to metres.

Q2 A jet-ski starts from the origin with a velocity of $(4\mathbf{i} + 6\mathbf{j}) \text{ m s}^{-1}$. It accelerates at $(-\mathbf{i} + 2\mathbf{j}) \text{ m s}^{-2}$, for 5 seconds. The unit vectors **i** and **j** are directed east and north respectively.

(a) Find the velocity and speed of the jet-ski after 5 seconds.

(b) Find the distance between the initial and final positions of the jet-ski.

(c) Find the average velocity of the jet-ski over the 5-second period.

(d) Find the time at which the jet-ski is heading due north.

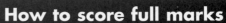

(a) $\underline{v} = (4\underline{i} + 6\underline{j}) + (-\underline{i} + 2\underline{j}) \times 5$

$\qquad = -\underline{i} + 16\underline{j}$ m s^{-1}

$\qquad v = \sqrt{1^2 + 16^2} = 16.03$ m s^{-1} (4 s.f.)

(b) $\underline{r} = (4\underline{i} + 6\underline{j}) \times 5 + \frac{1}{2} \times (-\underline{i} + 2\underline{j}) \times 5^2$

$\qquad = 7.5\underline{i} + 55\underline{j}$

$\qquad d = \sqrt{7.5^2 + 55^2} = 55.5$ m (3 s.f.)

(c) Average velocity $= \dfrac{7.5\underline{i} + 55\underline{j}}{5}$

$\qquad\qquad\qquad\quad = 1.5\underline{i} + 11\underline{j}$

(d) $\underline{v} = (4\underline{i} + 6\underline{j}) + (-\underline{i} + 2\underline{j})t$

$\qquad = (4 - t)\underline{i} + (6 + 2t)\underline{j}$

If the jet-ski is travelling due north, the velocity in the east-direction must be 0.

This means $4 - t = 0$

$\Rightarrow t = 4$

Q3 An aeroplane flies due north at a speed of 80 m s^{-1}. It flies through air that is moving north-east at 30 m s^{-1}. Find the resultant velocity of the aeroplane, expressing it as a speed and the bearing on which the aeroplane is actually moving.

$v^2 = 30^2 + 80^2 - 2 \times 30 \times 80 \times \cos 135°$

$v = 103$ m s^{-1} (3 s.f.)

The bearing will be α.

$\dfrac{\sin\alpha}{30} = \dfrac{\sin 135°}{103}$

$\alpha = 12°$ to the nearest degree

30 m s^{-1}

135°

resultant velocity

80 m s^{-1}

α

The aeroplane is travelling at 103 m s^{-1} on a bearing of 012°.

How to score full marks

- Use the motion equation $\mathbf{v} = \mathbf{u} + \mathbf{a}t$ to find the velocity.
- Remember that the speed is the **magnitude** of the velocity.
- Now you can find the final position, using $\mathbf{r} = \mathbf{u}t + \frac{1}{2}\mathbf{a}t^2$, then the magnitude of $\mathbf{r}$ gives the **distance**.

- The **average velocity** is the displacement, found in (b), divided by the time.

- The jet-ski will be heading north when the **i**-component is 0. You need to know this, then you can form the equation and solve it.

- By drawing a clear sketch, you can show the examiner that you understand the question and avoid making errors.
- Remember that the speed is represented by the length (or **magnitude**) of the **resultant velocity** vector.
- You need to remember both the **cosine rule** and the **sine rule** for this type of problem. In this example, you use the cosine rule first, to find the **speed** (a length), and then the sine rule to find the **bearing** (an angle).

Questions to try

Q1 During a 10-second period the velocity of a boat changes from $(4\mathbf{i} + 2\mathbf{j})\,\text{m\,s}^{-1}$ to $(\mathbf{i} - 3\mathbf{j})\,\text{m\,s}^{-1}$, where $\mathbf{i}$ and $\mathbf{j}$ are perpendicular unit vectors. Find the acceleration of the boat during this time, assuming that it is constant.

Q2 An object has initial velocity $(5\mathbf{i} - 5\mathbf{j})\,\text{m\,s}^{-1}$ and an acceleration of $(-\mathbf{i} + 2\mathbf{j})\,\text{m\,s}^{-2}$, where $\mathbf{i}$ and $\mathbf{j}$ are unit vectors directed east and north respectively.

   (a) If the object starts at the origin, find expressions for its position and velocity after t seconds.
   (b) Find the time at which the object is travelling due north.

Q3 Two model boats on a pond are set into motion. The unit vectors $\mathbf{i}$ and $\mathbf{j}$ are east and north respectively. Boat A has a constant velocity of $(4\mathbf{i} + 3\mathbf{j})\,\text{m\,s}^{-1}$ and starts at a point with position vector $4\mathbf{j}$. Boat B has a constant velocity of $(2\mathbf{i} - \mathbf{j})\,\text{m\,s}^{-1}$ and initial position $(4\mathbf{i} + 12\mathbf{j})\,\text{m}$.

   (a) Find the position vector of each boat at time t seconds.
   (b) Show that the boats collide and find the position of the boats at this time.
   (c) Find the distance between the two boats when $t = 1$.

Q4 A swimmer can move through still water at $1.2\,\text{m\,s}^{-1}$. She swims in a straight line in a river flowing at $0.8\,\text{m\,s}^{-1}$. She travels from the point A to the point B, so that her resultant velocity makes an angle of $30°$ to the downstream bank, as shown in the diagram.

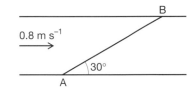

   (a) Sketch an appropriate triangle of velocities.
   (b) Use either scale drawing or trigonometry to find the magnitude of her resultant velocity.

Q5 A model assumes that an aeroplane has an initial velocity of $200\mathbf{i}\,\text{m\,s}^{-1}$ and experiences an acceleration of $(-0.5\mathbf{i} - 0.05\mathbf{j})\,\text{m\,s}^{-2}$ in preparation for landing. The initial position of the aeroplane was $(-50\,000\mathbf{i} + 4000\mathbf{j})\,\text{m}$, with respect to an origin O. The unit vectors $\mathbf{i}$ and $\mathbf{j}$ are horizontal and vertical respectively. After the aeroplane has accelerated for 200 seconds, its velocity is assumed to remain constant until it lands. The aeroplane lands when the vertical component of its position vector is zero.

   (a) Find: **(i)** the time it takes for the aeroplane to move from its initial position to the point where it first touches the ground
           (ii) the position of the aeroplane when it first touches the ground
           (iii) the speed of the aeroplane when it first touches the ground.

   (b) Comment on the assumption that the velocity is constant during the final stage of the aeroplane's flight.

Q6 Two cars A and B are moving on straight, horizontal roads with constant velocities. The velocity of A is $20\,\text{m\,s}^{-1}$ due east, and the velocity of B is $(10\mathbf{i} + 10\mathbf{j})\,\text{m\,s}^{-1}$, where $\mathbf{i}$ and $\mathbf{j}$ are unit vectors directed due east and due north respectively. Initially A is at the fixed origin O, and the position vector of B is $300\mathbf{i}\,\text{m}$ relative to O. At time t seconds, the position vectors of A and B are $\mathbf{r}$ and $\mathbf{s}$ metres respectively.

   (a) Find expressions for $\mathbf{r}$ and $\mathbf{s}$ in terms of t.
   (b) Hence write down an expression for AB in terms of t.
   (c) Find the time when the bearing of B from A is $045°$.
   (d) Find the time when the cars are $300\,\text{m}$ apart again.

Answers can be found on pages 119–120.

Key points to remember

- Identify the **different forces acting on a particle** and draw clear force diagrams showing these forces.

- The common forces that occur in most problems are **gravity**, **friction**, **normal reaction**, **tension in a string** and the **tension or compression in a rod**.

- In **equilibrium** a particle is at rest or moving with constant velocity. There is **no change in motion**.

- In **equilibrium** the **resultant force is zero** and the **resultant moment is zero**. If the particle is not at rest or moving with constant velocity then Newton's second law is the equation of motion.

- For problems involving connected particles, apply **Newton's second law to each particle** in the system separately.

Formulae you must know

- The law of friction: $F \leqslant \mu R$
  where F is the friction, μ is the coefficient of friction and R is the normal reaction

- Newton's second law: resultant force = mass × acceleration

- For a particle to remain at rest, the resultant force must be zero

- 1 tonne ≡ 1000 kg

Don't make these mistakes...

Don't always assume that the **normal reaction** is equal to the weight.

Take care when **resolving** – don't use the wrong angle.

When drawing or labelling a **force diagram**, only include **forces that actually exist**.

Don't forget to include **negative signs** for forces that act to **oppose the motion**.

Q1

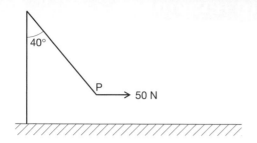

A bowling ball P is attached to one end of a light, inextensible string, the other end of the string being attached to the top of a fixed vertical pole. A girl applies a horizontal force of magnitude 50 N to P, and P is in equilibrium under gravity with the string making an angle of 40° with the pole, as shown in the diagram.

By modelling the ball as a particle, find, to three significant figures,

(a) the tension in the string
(b) the weight of P.

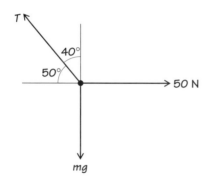

- You must always start with a clear force diagram showing the forces on the ball. This shows the examiner that you understand the question, and helps you to avoid making mistakes.

(a) Resolving the forces horizontally:

$T\cos 50° = 50$

$T = 77.8\,N$

(b) Resolving the forces vertically:

$T\cos 40° = mg$ (the weight)

$mg = 59.6\,N$

The weight of the ball is 59.6 N.

- Since the bowling ball is held at rest the components of the forces must balance in any direction.

- In this example, it is sensible to resolve the forces into horizontal and vertical components.

Q2

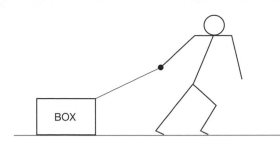

The diagram shows a box of weight 150 newtons which is being pulled along a rough horizontal surface at constant speed. The tension in the rope pulling the box is 50 newtons, and the rope makes an angle of 30° with the horizontal.

The coefficient of friction between the box and the surface is μ.

(a) Draw a diagram showing the forces acting on the box.
(b) Calculate the value of:
  (i) the normal reaction force R
  (ii) the friction force F.
(c) Calculate the value of μ.

(a)

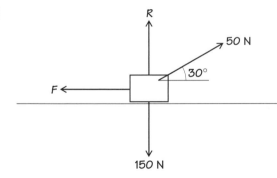

- The forces are the normal reaction R, friction force F, the weight and the tension in the string.

(b) (i) Resolving the forces vertically:

$R + 50\cos60° = 150$

$R = 125$ N

- The box is moving with constant speed along the surface, so the forces are in balance.

(ii) Resolving the forces horizontally:

$F = 50\cos 30° = 25\sqrt{3}$ N

(c) Applying the law of friction:

$F = \mu R \Rightarrow \mu = \dfrac{25\sqrt{3}}{125} = \dfrac{\sqrt{3}}{5}$

- Remember that, **since the box is sliding**, the friction law is:

$$F = \mu R$$

The diagram shows a railway engine of mass 50 tonnes pulling two trucks horizontally along a straight track. The trucks are coupled together behind the engine and have masses 8 tonnes and 4 tonnes respectively, starting with the truck nearer to the engine. The acceleration of the train is $0.5\,\mathrm{m\,s^{-2}}$.

Assuming that there are no resistances to motion, find:

(i) the driving force of the engine
(ii) the tensions in the two couplings.

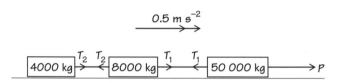

(i) Applying Newton's second law to the whole system of engine and trucks:

$$P = (50\,000 + 8000 + 4000)a$$

$$P = 31\,000\ \text{N}$$

(ii) Applying Newton's second law to the engine and each truck separately:

Engine: $\quad P - T_1 = 50\,000 \times \frac{1}{2}$

$$T_1 = 31\,000 - 25\,000$$

$$T_1 = 6000\ \text{N}$$

8000-kg truck: $\quad T_1 - T_2 = 8000 \times \frac{1}{2}$

$$T_2 = 6000 - 4000$$

$$T_2 = 2000\ \text{N}$$

- First, you need to draw a force diagram showing the forces between the engine and trucks.

- Remember that 1 tonne = 1000 kg

- For all problems like this one, you need to apply Newton's second law to each object in the system to find the separate forces. If you apply Newton's second law to the whole system, remember to add all the masses to get the total mass of the system.

- Remember that, for the whole system, the internal forces (T_1 and T_2) cancel out.

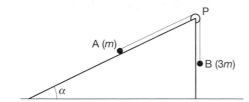

The particle, A, of mass m rests on a rough plane inclined at an angle α to the horizontal, where $\tan\alpha = \frac{3}{4}$.

The particle is attached to one end of a light, inextensible string which passes over a small, light smooth pulley P fixed at the top of the plane. The other end of the string is attached to a particle B of mass $3m$, and B hangs freely below P. The particles are released from rest with the string taut. The particle B moves down with acceleration of magnitude $\frac{1}{2}g$. Find:

(a) the tension in the string

(b) the coefficient of friction between A and the plane.

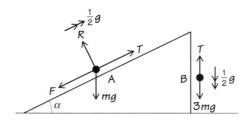

● Start by writing the forces on the diagram. You might also find it helpful to write down useful information, such as the trigonometric ratios, as this student has done.

$\tan\alpha = \frac{3}{4} \quad \cos\alpha = \frac{4}{5} \quad \sin\alpha = \frac{3}{5}$

(a) Applying Newton's second law to B:

$3mg - T = 3m \times \frac{1}{2}g \Rightarrow T = \frac{3}{2}mg$

(b) Apply Newton's second law to A.

● Newton's second law is the equation of motion.

Perpendicular to plane:
$R - mg\cos\alpha = 0 \Rightarrow R = \frac{4}{5}mg$

Parallel to plane:
$T - mg\sin\alpha - F = m \times \frac{1}{2}g$

● Remember that friction F opposes the direction of motion of A.

$F = T - mg\sin\alpha - \frac{1}{2}mg$

$F = \frac{3}{2}mg - \frac{3}{5}mg - \frac{1}{2}mg = \frac{2}{5}mg$

The law of sliding friction gives $F = \mu R$.

● Remember that $F \leqslant \mu R$ if the object is at rest. Otherwise $F = \mu R$ when the object is sliding or about to slide.

Coefficient of friction $\mu = \dfrac{F}{R} = \dfrac{\frac{2}{5}mg}{\frac{4}{5}mg} = \dfrac{1}{2}$

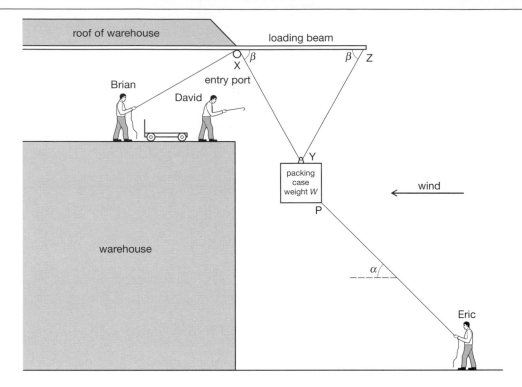

The diagram shows three men involved in hoisting a packing case up to the top floor of a warehouse. Brian is pulling on a rope which passes round smooth pulleys at X and Y and is then secured to the point Z at the end of the loading beam.

The wind is blowing directly towards the building. To counteract it, Eric is pulling on another rope, attached to the packing case at P, with just enough force and in the right direction to keep the packing case central between X and Z.

At the time of the sketch, the men are holding the packing case motionless.

(i) Draw a diagram showing and labelling all the forces acting on the packing case. Treat the pulley Y as part of the packing case.
Use W N for the weight of the packing case
$\quad$ T_1 N for the tension in Brian's rope
$\quad$ T_2 N for the tension in Eric's rope
and F N for the force of the wind on the packing case.

(ii) If Brian had positioned himself further away from the entry port, but had used more rope so the position of the packing case stayed the same, would the tension in his rope have been greater, the same or less? Explain your answer briefly.

(iii) Write down equations for the horizontal and vertical equilibrium of the packing case. (Notice the angles α and β in the sketch.)

In one particular situation it is found that $W = 200$, $F = 50$, $\alpha = 45°$ and $\beta = 75°$.

(iv) Find the tension T_1.

(v) Given that Brian's rope is at 30° to the roof of the warehouse, calculate the magnitude of the force acting on the pulley at X due to the tension in the rope.

(vi) Explain why Brian has to pull harder if the wind blows stronger.

Q2

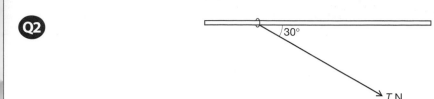

A heavy ring of mass 5 kg is threaded on a fixed rough horizontal rod. The coefficient of friction between the ring and the rod is $\frac{1}{2}$. A light string is attached to the ring and is pulled with a force of magnitude T newtons acting at an angle of 30° to the horizontal (see diagram). Given that the ring is about to slip along the rod, find the value of T.

Q3 A block, of mass 6 kg, rests on a rough horizontal surface. The coefficient of friction between the block and the surface is 0.2. A light, inextensible string attached to the block passes over a smooth pulley. A weight, of mass 2 kg, hangs from the other end of the string, as shown in the diagram below.

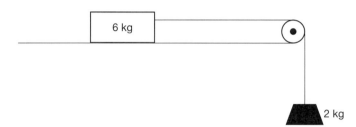

Find the tension in the string and the acceleration of the block.

Q4

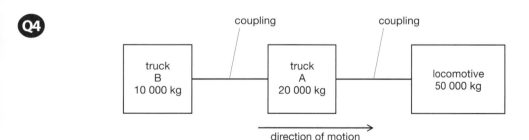

A train made up of a locomotive and two trucks is travelling along a straight, level track. There is a driving force of 10 000 N from the locomotive. The resistances to forward motion are 500 N on each of the trucks and 1000 N on the locomotive. The masses of the locomotive and the two trucks are shown in the diagram.

(i) Calculate the acceleration of the train and the forces in each of the two couplings.

With the first locomotive still exerting a driving force of 10 000 N, a second locomotive is added to the train. This locomotive is behind truck B and has the effect of applying a forward force of 4000 N to it.

(ii) Show that the coupling between trucks A and B is now under a compression of 2000 N.

The second locomotive has a mass of 40 000 kg and resistance to forward motion of 800 N.

(iii) Calculate the total driving force of the second locomotive.

Q5 Two light scale pans, A and B, are connected by a light cord which hangs over a smooth peg. Two particles of mass pm and qm are placed on A and B, respectively, and the system is released from rest. In the resulting motion, A moves downwards with acceleration $\frac{1}{2}g$ or $\frac{g}{2}$.

(a) (i) Find, in terms of p, m and g, the tension in the string.
  (ii) Show that $p = 3q$.

(b) The particles of mass pm and qm are then both placed on the scale pan A and a particle of mass $4m$ is placed on B. The system is again released from rest and A again accelerates downwards at a rate of $\frac{1}{2}g$ or $\frac{g}{2}$.

  (i) By using the result in (a)(ii), or otherwise, find another equation in p and q.
  (ii) Hence find the values of p and q.

Answers can be found on pages 121–123.

Key points to remember

- **Momentum** is the product of the **mass and velocity** of a body.
- In collisions, **momentum is conserved** if no external forces act.
- The change in momentum of a body is the **impulse**.
- When a **constant force** acts, the impulse is equal to the product of the force and the time for which it acts.

Formulae you must know

- $m_A u_A + m_B u_B = m_A v_A + m_B v_B$
- $I = mv - mu$
- $I = Ft$

Don't make these mistakes...

Don't forget to **include the signs** when working with negative velocities.

Don't forget to define a positive direction, when starting a problem.

Exam Questions and Student's Answers

Q1 Two particles are moving towards each other. Particle A has mass 2 kg and speed $3\,\text{m}\,\text{s}^{-1}$. Particle B has mass 4 kg and speed $5\,\text{m}\,\text{s}^{-1}$. After the collision particle B moves in the same direction, but its speed has been reduced to $2\,\text{m}\,\text{s}^{-1}$. Describe the motion of particle A after the collision.

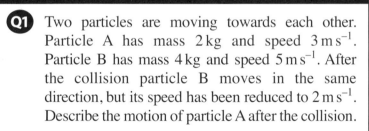

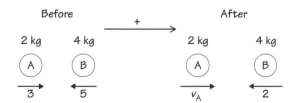

Using conservation of momentum:

$2 \times 3 + 4 \times (-5) = 2v_A + 4 \times (-2)$

$$-14 = 2v_A - 8$$

$$v_A = -3$$

So after the collision particle A has changed direction and is moving at $3\,\text{m}\,\text{s}^{-1}$.

How to score full marks

- The student has drawn a diagram to show the velocities of the particles before and after the collision. This is a very good way to start problems like this one, as it gives a summary of the information you need.

  Note that:
  $m_A = 2 \quad u_A = 3 \quad v_B = -2$
  $m_B = 4 \quad u_B = -5$

- You can substitute these values into the equation for the conservation of momentum.

Q2 A train carriage, of mass 50 tonnes, is travelling at $10\,\text{m s}^{-1}$, when it collides with a second carriage, of mass 100 tonnes, travelling at $6\,\text{m s}^{-1}$ in the same direction. After the collision the two carriages travel together at the same speed.

Find the speed of the carriages after the collision.

Before   After

50 T 100 T   150 T

| A | B | | A | B |

$\underrightarrow{10}$ $\underrightarrow{6}$ $\underrightarrow{v}$

Using conservation of momentum:

$$50\,000 \times 10 + 100\,000 \times 6 = 150\,000v$$

$$1\,100\,000 = 150\,000v$$

$$v = 7.33 \ (3\ \text{s.f.})$$

The speed after the collision is $7.33\,\text{m s}^{-1}$.

- Again, the student has started with a diagram to show the information given in the question.

- The diagram shows the masses and velocities of the carriages before and after the collision.

- Note that after the collision the two carriages can be considered as single particle. In this case:

$m_A = 50\,000$ $u_A = 10$ $v_A = v_B = v$
$m_B = 100\,000$ $u_B = 6$

- If you make a list of these when you start, then you can substitute them into the conservation of momentum equation.

Q3 A ball, of mass 300 grams, bounces on a horizontal surface. When it hits the ground it is travelling vertically downwards at $5\,\text{m s}^{-1}$ and it rebounds vertically at $3\,\text{m s}^{-1}$.

(a) Calculate the magnitude of the impulse on the ball.

(b) If the ball is in contact with the ground for 0.2 seconds, find the average magnitude of the resultant force on the ball.

(a) Take the downward direction as positive.

$$I = mv - mu$$

$$= 0.3 \times (-3) - 0.3 \times 5$$

$$= -2.4\,\text{Ns}$$

The magnitude of the impulse is 2.4 Ns.

(b) As the contact time is 0.2 seconds the average force can be found using

$$I = Ft.$$

$$2.4 = F \times 0.2$$

$$F = 12\,\text{N}$$

- It is a good start to state which direction you are taking as positive, as this student has done.

- As the downward direction is being taken as positive, the upward direction is negative, and $u = 5$ and $v = -3$.

- You can find the impulse by considering the change in momentum.

- The impulse is negative, but as you want the magnitude, you just remove the minus sign.

- Again, as you have been asked for the magnitude of the force, you can use the magnitude of the impulse.

Questions to try

Q1 A particle, of mass 0.5 kg, is at rest when it is hit by another particle, moving at $3\,\mathrm{m\,s^{-1}}$. After the collision the two particles move together at $1.4\,\mathrm{m\,s^{-1}}$.

 (a) Find the mass of the particle that was initially moving.
 (b) Find the impulse on the particle that was initially at rest.

Q2 A train, of mass 200 tonnes, is moving at $0.8\,\mathrm{m\,s^{-1}}$, when it hits a stationary carriage of mass 20 tonnes. After the collision the velocity of the train is reduced to $0.7\,\mathrm{m\,s^{-1}}$.

 (a) Find the speed of the carriage after the collision.
 (b) Find the impulse on the train during the collision.
 (c) If the trains are in contact for 0.5 seconds, find the average force exerted by the carriage on the train.

Q3 A child, of mass 48 kg, jumps off a stationary skateboard. He initially moves horizontally at $0.5\,\mathrm{m\,s^{-1}}$. The skateboard has mass 1.2 kg. Calculate the initial speed of the skateboard.

Q4

Two particles, of masses x kg and 0.1 kg, are moving towards each other in the same straight line and collide directly. Immediately before the impact, the speeds of the particles are $2\,\mathrm{m\,s^{-1}}$ and $3\,\mathrm{m\,s^{-1}}$ respectively (see diagram).

 (a) Given that both particles are brought to rest by the collision, find x.
 (b) Given instead that the particles move with equal speeds of $1\,\mathrm{m\,s^{-1}}$ after the impact, find the three possible values of x.

Q5 A ball, of mass m kg, is dropped from a height of h m, so that it falls vertically to the floor. It hits the floor and rebounds to a height of H m.

 (a) Find the speed at which the ball hits the floor and the speed at which it rebounds.
 (b) Find the magnitude of the impulse on the ball.

Q6 Train A has mass 3 tonnes and moves on straight horizontal rails. It collides with train B, which has a mass of 1 tonne and is moving on the same rails. Immediately before the collision both trains were moving towards each other. The speed of A was $3\,\mathrm{m\,s^{-1}}$ and the speed of B was $1.5\,\mathrm{m\,s^{-1}}$. During the collision the trains couple together and form a single body.

 (a) Find the speed of the two combined trains after the collision and state which train has changed its direction of motion.
 (b) Calculate the magnitude of the impulse on train A.

Answers can be found on pages 123–124.

Key points to remember

- The set of all data being considered is called the **population**.

- A **sample** is part of the population.

- Nearly all real data is a sample because it is too difficult, time-consuming or expensive to collect the whole population.

- A sample is taken to obtain information about the population.

- A numerical property of a population (such as the arithmetic mean) is a **parameter**.

- A numerical property of a sample is a **statistic**.

- Every member of the population must have an equal chance of being included in a random sample. For a random sample of size n, all subsets, of size n, of the population must have equal chances of forming the sample.

- Random samples are usually selected using **random numbers** to remove the human element from the selection.

- In a **stratified sample**, prior knowledge of the population is used to ensure that different strata (such as male and female) are represented fairly in the sample.

- A **quota sample** is a stratified sample where the choice of the individual members of the population is left to the interviewer. It is used when taking a random sample from each stratum is impractical for reasons of time, expense … .

- **Cluster sampling** is used where random sampling would be impractical, usually for geographical reasons. For example, to interview a random sample of all junior school teachers in the United Kingdom would involve excessive travelling. To reduce this a random sample of junior schools could be selected and all the teachers in these schools would be interviewed.

- In a **systematic sample**, every 100th name on an electoral register, for example, or every 15th car off a production line, is selected.

Don't make these mistakes ...

Don't assume that all members of the population having an equal chance of being selected is a sufficient condition to obtain a random sample. This may be true of stratified, systematic and in some circumstances cluster samples but **none of these are random samples**.

Q1 Describe how random numbers could be used to select a simple random sample (i.e. without replacement) of size six from the 712 employees of a large city centre store.

> Number the employees from 000 to 711.
>
> Select three-digit random numbers.
>
> Ignore repeats. Ignore 712 and over.
>
> Continue until six numbers have been obtained. Choose corresponding employees.

- It is possible but unusual to take a random sample with replacement. This is called an **unrestricted random sample**.

- It is also acceptable – but more complicated – to number from 001 to 712. You will have to ignore 000 and 713 and over.

Q2 Explain briefly what you understand by a quota sample. Under what circumstances would you advise the use of quota sampling?

> Quota sampling is common when carrying out surveys. Prior knowledge of the population is used to divide it into strata. Interviewers are told to select a given number of people in a given stratum, for example, 10 females, aged over 60, living in an inner city area. The choice of people within the strata is left to the interviewer. There is no attempt at random sampling. It should be used when relevant prior knowledge of the population is available but random sampling from the strata is too difficult/expensive/time consuming.

- The question does not ask for an example but giving one is a good way of clarifying your answer.

Q3 In a particular parliamentary constituency there are 64 000 names on the electoral register. Of the electors, 32 000 live in property rented from the local authority, 21 000 live in owner-occupied property and 11 000 live in other types of property.

The following methods are suggested for choosing a sample of electors in order to carry out an opinion survey.

A Use a random process to select 128 names from the electoral register.

B Use a random process to select one of the first 500 names on the electoral register. Using this as a starting point select every 500th name.

C Select 64 names at random from the electors living in property rented from the local authority, 42 names at random from the electors living in owner-occupied property and 22 names at random from those living in other types of property.

(a) For each of the methods A, B and C:
   (i) name the type of sampling method
   (ii) state whether all the names on the electoral register are equally likely to be included in the sample.

(b) State, giving a reason, whether method C will produce a random sample of the electors.

(c) State, briefly, the difference between a sample obtained by method C and a quota sample. What is the advantage of a quota sample compared to a sample obtained by method C?

(d) Compare the usefulness of sampling methods A and C if the questions to be asked concerned local authority housing policy.

(e) How would your answer to part (d) be changed (if at all) if the questions to be asked concerned attitudes to the monarchy? Explain your answer.

(a) (i) A random sampling
     B systematic sampling
     C stratified sampling
  (ii) All are equally likely.

(b) C is not a random sample since not all combinations of electors can be chosen.

(c) In a quota sample the electors are not selected from the strata at random. Quota sampling is easier/quicker/cheaper.

(d) Stratification is directly relevant to the survey. Different strata will almost certainly differ in their views of housing policy, so C will be better than A as it will ensure that each stratum is fairly represented.

(e) There is no obvious reason why the strata should differ in their views on the monarchy (although they may). If they do not, there is no advantage in using method C. The only disadvantage is that it involves more work than A.

- It wouldn't be equally likely if 64 000 was not exactly divisible by 500.

- Sometimes called 'stratified random sampling' to distinguish it from quota sampling which is also a form of stratified sampling.

Q1 A warehouse contains 589 TV sets. Describe how random numbers could be used to select a random sample of 12 TV sets from the warehouse.

Q2 Explain, briefly, what you understand by a **stratified** sample. Under what circumstances would you recommend the use of stratified sampling?

Q3 Explain briefly what is meant by a **systematic** sample. Give one advantage and one disadvantage of systematic sampling compared to random sampling.

Q4 Describe, with the aid of an example, the meaning of **cluster** sampling. Give one advantage and one disadvantage compared to random sampling.

Q5 There are 28 houses in Mandela Road, 14 on each side. The houses on one side of the road have even numbers and those on the other side of the road have odd numbers.

A total of 63 residents of Mandela Road are on the electoral register.

A market researcher wishes to interview seven of these residents. He decides to choose a sample of seven houses using the following procedure:

Step 1 Toss a coin and choose the side of the road with odd-numbered houses if it falls heads and the side of the road with even-numbered houses if it falls tails.

Step 2 Toss the coin again and select the lowest-numbered house on the chosen side of the road if it falls heads and the second lowest if it falls tails.

Step 3 Select alternate houses on the chosen side of the road starting from the house chosen in step 2.

(For example, a tail followed by another tail would result in him selecting the houses numbered 4, 8, 12, 16, 20, 24 and 28.)

(a) (i) Would all houses in Mandela Road be equally likely to be included in the sample? Explain your answer.
    (ii) Would the sample be random? Give a reason for your answer.

The market researcher knocks at each selected house and asks the person who opens the door to answer a questionnaire. Assume that each person who opens the door is on the electoral register and is willing to answer the questionnaire.

(b) Give **two** reasons why the people who answer the questionnaire are not a random sample from the 63 residents on the electoral register.

(c) Describe how random numbers could be used to select a random sample of size seven (without replacement) from the 63 residents on the electoral register.

Q6 A manufacturing company operates 15 factories of varying size. It has a total of 13 800 employees. The board of directors, concerned by a large turnover of staff, decides to survey 100 employees to seek their opinions on working conditions.

The following suggestions were made as to how the sample could be chosen.

Suggestion A: The employees are numbered 00000 to 13799. One hundred different five-digit random numbers between 00000 and 13799 are taken from random number tables and the corresponding employees chosen.

Suggestion B: The sample is made up of employees from all factories. The employees are selected at random from each factory, the number from each factory being proportional to the number of employees at the factory.

Suggestion C: The sample is made up of employees from all factories. The employees are selected by any convenient method, the number from each factory being proportional to the number of employees at the factory.

(a) (i) Which of the suggestions would produce a **quota** sample?
    (ii) Name the type of sampling described in each of the other two suggestions.
(b) For each of the suggestions state whether or not all employees have an equal chance of being included in the sample.
(c) Give one reason for using:
    (i) **Suggestion A** in preference to **Suggestion C**
    (ii) **Suggestion C** in preference to **Suggestion A**.
(d) Explain why **Suggestion B** might be preferred to **Suggestion A**.

Answers can be found on pages 124–125.

18 Probability

Key points to remember

- If a trial can result in a number of possible **outcomes**, each outcome may have a probability associated with it.

- Probability is measured on a scale from 0 to 1. 0 represents **impossibility** and 1 represents **certainty**.

- If an outcome has probability 0.2 it is expected to occur, in the long run, in 0.2 or $\frac{1}{5}$ of all trials.

- **Events** consist of **one or more outcomes**.

- Examination questions will either give you the probability of an event, or require you to derive it using '**equally likely outcomes**'.

- The event that A does not occur is usually denoted A'. $P(A') = 1 - P(A)$.

- Two events are **mutually exclusive** if they cannot both occur as a result of the same trial.

- If A and B are mutually exclusive events **P(A∪B) = P(A) + P(B)**. P(A∪B) represents the probability of event A **or** event B happening. It includes the case of both happening (but this cannot occur with mutually exclusive events).

- The law may be extended to more than two mutually exclusive events.

- A and B are **independent events** if the probability of A happening is not affected by whether B happens. Independent events could be different outcomes of the same trial but are much more likely to be outcomes of different trials.

- P(A | B) is the **conditional probability** of A happening given that B happens. A and B may occur simultaneously but it is easier to think of A happening after B.

- If A and B are independent then **P(A | B) = P(A)**. It is also true that if P(A | B) = P(A) then A and B are independent.

- **P(A∩B) = P(A) × P(B | A)**.

- P(A∩B) represents the probability of events A and B both occurring. If A and B are **independent**, this law simplifies to **P(A∩B) = P(A) × P(B)**.

- This law may be extended to more than two events.

Formulae you must know

- $P(A') = 1 - P(A)$

- $P(A∪B) = P(A) + P(B)$ provided A and B are mutually exclusive.

  This may be extended to $P(A∪B∪C…) = P(A) + P(B) + P(C) + …$ provided A, B, C … are mutually exclusive.

- $P(A∩B) = P(A) × P(B | A)$

  This may be extended to $P(A∩B∩C…) = P(A) × P(B | A) × P(C | A \text{ and } B) × …$

  If A, B, C … are independent this simplifies to:

  $P(A∩B∩C…) = P(A) × P(B) × P(C)…$

Don't make these mistakes ...

Don't use a probability that is **negative or greater than 1**. If your calculations produce probabilities like this, **you have definitely made a mistake**.

When calculating the probability of **two events both occurring** don't add the probabilities. **Multiply** them – the probability of two events both occurring will be less than their individual probabilities of occurring.

When calculating the probability of **one or other of two events occurring**, don't multiply the probabilities. **Add** them – the probability of one or other event occurring will be greater than their individual probabilities of occurring.

Q1 The probability that callers to a railway timetable enquiry service receive accurate information is 0.9. Find the probability that of three randomly selected callers:

(a) all receive accurate information
(b) exactly two receive accurate information
(c) less than two receive accurate information.

(a) $0.9 \times 0.9 \times 0.9 = 0.729$

(b) $P(A) = 0.9 \quad P(I) = 1 - 0.9 = 0.1$

$\quad$ AAI: $\quad 0.9 \times 0.9 \times 0.1 = 0.081$
$\quad$ AIA: $\quad 0.9 \times 0.1 \times 0.9 = 0.081$
$\quad$ IAA: $\quad 0.1 \times 0.9 \times 0.9 = 0.081$

$\quad$ P(2 customers receiving accurate information) $= 3 \times 0.081 = 0.243$

(c) P($\geqslant$2 customers receiving accurate information)
$\quad = 0.729 + 0.243 = 0.972$

$\quad$ P(<2 customers receiving accurate information)
$\quad = 1 - 0.972 = 0.028$

- Each customer will receive either accurate or inaccurate information.
- All three must receive accurate information. Events are independent. Multiply the individual probabilities.
- Each line is the probability of a particular customer receiving inaccurate information and the other two receiving accurate information. There are three mutually exclusive ways in which this can happen, so you need to add the probabilities.
- You could calculate P(less than 2) directly but it is easier to use the results you have already found for P(not less than 2).
- Alternatively, you could use a tree diagram to show all the possible outcomes and their probabilities. You may find this method easier but it only really works for a relatively small number of alternatives.

Q2 Three persons are to be selected at random, one after another, from a group of eight persons of whom five are female and three are male. Calculate the probability that:

(a) each of the first two persons selected will be female and the third will be male
(b) two females and one male will be selected
(c) all three selected will be of the same sex.

(a) $\frac{5}{8} \times \frac{4}{7} \times \frac{3}{6} = 0.179$

(b) FFM: $\frac{5}{8} \times \frac{4}{7} \times \frac{3}{6} = \frac{5}{28}$

$\quad$ FMF: $\frac{5}{8} \times \frac{3}{7} \times \frac{4}{6} = \frac{5}{28}$

$\quad$ MFF: $\frac{3}{8} \times \frac{5}{7} \times \frac{4}{6} = \frac{5}{28}$

$\quad$ P(2 females and 1 male) $= 3 \times \frac{5}{28} = 0.536$

(c) FFF: $\frac{5}{8} \times \frac{4}{7} \times \frac{3}{6} = \frac{5}{28}$

$\quad$ MMM: $\frac{3}{8} \times \frac{2}{7} \times \frac{1}{6} = \frac{1}{56}$

$\quad$ P(all 3 the same sex) $= \frac{5}{28} + \frac{1}{56} = 0.196$

- This question is about conditional probability. If the first person selected is female, that leaves seven, of whom four are female. After two females have been selected there are six left, of whom three are male.

- Note that although the sexes are selected in different orders the probabilities of each order are the same.
- There are three mutually exclusive ways of selecting two females and one male. You need to add the probabilities.
- You could also have answered this question using a tree diagram.

Q3 Last year the employees of a firm received no pay rise, a small pay rise or a large pay rise. The following table shows the number in each category, classified by whether they were weekly paid or monthly paid.

	No pay rise	Small pay rise	Large pay rise
Weekly paid	25	85	5
Monthly paid	4	8	23

A tax inspector decides to investigate the tax affairs of an employee selected at random.

D is the event that a weekly employee is selected.

E is the event that an employee who received no pay rise is selected.

D' and E' are the events 'not D' and 'not E' respectively.

Find:

(a) P(D)   **(b)** P(D∪E)   **(c)** P(D'∩E').

F is the event that an employee is female.

(d) Given that P(F') = 0.8, find the number of female employees.

(e) Interpret P(D | F) in the context of this question.

(f) Given that P(D∩F) = 0.1, find P(D | F).

(a) $P(D) = \frac{115}{150} = 0.767$

(b) $P(D\cup E) = \frac{119}{150} = 0.793$

(c) $P(D'\cap E') = \frac{31}{150} = 0.207$

(d) $P(F) = 1 - 0.8 = 0.2$

There are $150 \times 0.2 = 30$ female employees.

(e) P(D | F) is the probability the employee selected is weekly paid given that she is female.

(f) $P(D\cap F) = P(F) \times P(D | F)$

$0.1 = 0.2 \times P(D | F)$

$P(D | F) = 0.5$

- There are 150 employees who are each equally likely to be selected. 115 of these are paid weekly.

- 119 employees are weekly paid or received no pay rise (or both).

- 31 employees are not paid weekly and received a pay rise (not 'no pay rise').

Q1 Customers at a supermarket pay by cash, cheque or credit card. The probability of a randomly selected customer paying by cash is 0.64 and by cheque is 0.13.

(a) Determine the probability of a randomly-selected customer paying by credit card.
(b) Two customers are selected at random. Find the probability of:
    (i) them both paying by credit card
    (ii) one paying by credit card and the other paying by cash.
(c) Three customers are selected at random. Find the probability of:
    (i) all three paying by cash
    (ii) exactly one paying by cheque
    (iii) one paying by cash, one by cheque and one by credit card.

Q2 When Bali is on holiday she intends to go for a five-mile run before breakfast each day. However, sometimes she stays in bed instead. The probability that she will go for a run on the first morning is 0.7. Thereafter, the probability that she will go for a run is 0.7 if she went for a run on the previous morning and 0.6 if she did not.

Find the probability that on the first three days of the holiday she will go for
(a) three runs
(b) exactly two runs.

Q3 The probability that a parachutist lands in a target area depends upon weather conditions.

If it is windy when he jumps, then the probability of landing in the target area is $\frac{2}{5}$ and if it is not windy the corresponding probability is $\frac{4}{5}$. The probability that it is windy on a random day in June is $\frac{1}{6}$.

(a) Calculate the probability that when the parachutist jumps on a random day in June he lands in the target area.
(b) Given that the parachutist landed in the target area on 15 June last year, find the probability that 15 June was a windy day.

Q4 A refreshment stall at a summer fair sells spring water, orange juice and lemonade. The probability that a randomly selected customer will choose spring water is 0.2, orange juice is 0.35 and lemonade is 0.45.

Find the probability that three randomly selected customers all choose:
(a) spring water
(b) the same drink
(c) spring water, given that they all choose the same drink.

Q5 One hundred and twenty students register for a foundation course. At the end of a year they are recorded as pass, fail or withdraw. A summary of the results, classified by age, is shown.

	Age (years)		
	< 20	20–25	> 25
Pass	47	20	13
Fail	17	3	1
Withdraw	11	7	1

A student is selected at random from the list of students who had registered for the course.

Q denotes the event that the age of the selected student is < 20 years.
R denotes the event that the selected student passed.
S denotes the event that the selected student failed.
(Q', R' and S' denote the events not Q, not R and not S respectively.)

Determine the value of:
(a) P(R)
(b) P(Q∩R)
(c) P(Q∪S')
(d) P(R | Q').

It is known that 45 out of the 120 students are female.

F denotes the event that the selected student is female.

(e) Find P(F∩R) given that F and R are independent.
(f) Find P(F∩Q) given that P(Q | F) = 0.8.
(g) Find P(S | F) given that P(S∩F) = 0.05.

Q6 It is known that 1% of the population suffers from a certain disease. A diagnostic test for the disease gives a positive response with probability 0.98 if the disease is present. If the disease is not present, the probability of a positive response is 0.005.

(a) A test is applied to a randomly selected person.
  (i) Show that the probability of this test giving a positive response is 0.014 75.
  (ii) Given that the test gave a positive response, calculate the probability that the person has the disease.
(b) A randomly-selected person is tested and a positive response is obtained. This person is tested again. Assuming that the tests are independent, calculate the probability that this second test will give a positive response.

Answers can be found on page 126.

Key points to remember

- The **binomial distribution** is used for trials that have **two possible outcomes**. For example a commuter train either arrives on time or is late.

- The outcomes may be called '**success**' and '**failure**'.

- If the probability of 'success' in each trial is p then the probability of exactly r 'successes' in n independent trials is:

$$\binom{n}{r}p^r(1-p)^{n-r} \quad \text{where} \quad \binom{n}{r} = \frac{n!}{r!(n-r)!}$$

  For the binomial distribution to apply the trials must be **independent** and p must be constant.

- Either outcome may be defined as 'success' but the usual convention is to call the outcome with the smaller probability 'success'. Thus p will be ≤ 0.5.

- The distribution is **discrete** and the possible outcomes are $0, 1, 2, \ldots, n$.

- The binomial distribution has **mean** np and **variance** $np(1-p)$.

- The **standard deviation** is $\sqrt{np(1-p)}$ and is a more useful measure of spread than the variance.

- Tables of the binomial distribution list the probabilities of r or fewer successes in n trials. Although you should use these tables where possible, not all values of n and p can be included.

- Some calculators will calculate binomial probabilities directly.

- The notation **B(10, 0.2)** denotes a binomial distribution with $n = 10$ and $p = 0.2$.

Don't make these mistakes ...

Don't assume that any question involving trials with two outcomes may be answered using the binomial distribution. The distribution also assumes:

- a fixed number of trials
- the trials are independent
- p is constant.

Formulae you must know

- $P(R = r) = \binom{n}{r}p^r(1-p)^{n-r}$

- $0! = 1$ and therefore $\binom{n}{n} = 1$ and $\binom{n}{0} = 1$

Q1 A pottery produces large quantities of drinking mugs and reckons that the probability of each mug being classed as 'seconds' is 0.12.

A random sample of six mugs is taken from the production. Find the probability that this sample will contain:

(**a**) exactly two 'seconds'
(**b**) two or fewer 'seconds'
(**c**) more than one 'second'.

(a) As this is a binomial distribution, with $n = 6$, $p = 0.12$:

$P(R = 2)$

$= \dfrac{6.5.4.3.2.1}{2.1.4.3.2.1} \times (0.12)^2 \times (1 - 0.12)^{6-2}$

$= 0.129\,53 = 0.130$

(b) $P(R \leqslant 2) = P(0) + P(1) + P(2)$

$= 0.88^6 + 6 \times 0.12 \times 0.88^5 + 0.129\,53$

$= 0.974$

(c) $P(R > 1) = 1 - P(0) - P(1)$

$\qquad = 1 - 0.88^6 - 6 \times 0.12 \times 0.88^5$

$\qquad = 0.156$

- There are two possible outcomes of each trial. The question implies that p is constant and the trials are independent.

- Your tables are very unlikely to include B(6, 0.12) so you will have to use the formula to evaluate the probabilities.

- It is sensible to give the answer to three significant figures, but to keep five for use in calculations later in the question.

- You have already calculated P(2) so you don't need to calculate it again.

- You can use three significant figures here, as you will not be using this result later.

- This is much quicker than calculating P(2) + P(3) + P(4) + P(5) + P(6).

Q2 The probability that any A-level candidate will be absent at the start of an examination is 0.01, independently of whether other candidates are absent. Find the probability that, of 50 A-level candidates, the number absent at the start of the examination will be:

(**a**) 3 or fewer   (**b**) exactly 3
(**c**) 3 or more   (**d**) more than 1.

(a) As this is a binomial distribution, with $n = 50$, $p = 0.01$:

$P(3 \text{ or fewer}) = 0.9984$

(b) $P(3) = P(3 \text{ or fewer}) - P(2 \text{ or fewer})$

$\qquad = 0.9984 - 0.9862 = 0.0122$

(c) $P(3 \text{ or more}) = 1 - P(2 \text{ or fewer})$

$\qquad\qquad = 1 - 0.9862 = 0.0138$

(d) $P(\text{more than } 1) = 1 - P(1 \text{ or fewer})$

$\qquad\qquad = 1 - 0.9106 = 0.0894$

- The question states that the two possible outcomes of each trial are independent.

- It is possible, but very time consuming to calculate the answers as in the previous question. You will be provided with tables of B(50, 0.01) so save time by using them.

- P(r or fewer) can be read directly from the tables.

- You could round to three significant figures but as the answer is so close to 1 it is sensible to give four significant figures.

- In order to use the tables you need to express the required probability in terms of 'P(r or fewer)'. Be particularly careful about the difference between 'r or more' and 'more than r'.

Q3 A cyclist has to make a right turn on a busy main road on the way to work each weekday morning. The probability that he is able to make the turn without stopping is 0.3.

(a) Find the probability that, on 20 consecutive weekday mornings:
  (i) he turns without having to stop, 4 or more times
  (ii) he turns without having to stop exactly five times.

The cyclist also makes the turn on Saturday and Sunday on the way to visit his invalid mother.

(b) State, giving a reason, whether or not it is likely that the following random variables may be modelled by a binomial distribution.
  (i) Y, the number of times the cyclist turns without having to stop on 20 consecutive days (i.e. including Saturday and Sunday)
  (ii) Z, the number of consecutive weekdays before the cyclist turns without having to stop on four occasions

(a) (i) As this is a binomial distribution with $n = 20$, $p = 0.0.3$:

$$P(4 \text{ or more}) = 1 - P(3 \text{ or fewer})$$

$$= 1 - 0.107$$

$$= 0.893$$

(ii) $P(5) = P(5 \text{ or fewer}) - P(4 \text{ or fewer})$

$$= 0.4164 - 0.2375$$

$$= 0.179$$

(b) (i) This is not a binomial distribution as p is not constant.

(ii) This is not a binomial distribution, as n is not constant.

- There are two possible outcomes of each trial.
- The numbers of turns on different days are independent.
- You will have tables of B(20, 0.03), so you can save time by using them.

- There are two possible outcomes but the probability of having to stop will be different at weekends than on weekdays.
- The trials continue until the cyclist turns without having to stop 4 times. For a binomial the number of trials must be known in advance.

Questions to try

Q1 Copies of an advertisement for a course in practical statistics are sent to mathematics teachers in a large city. For each teacher who receives a copy, the probability of subsequently attending the course is 0.09.

Twenty teachers receive a copy of the advertisement. What is the probability that the number who subsequently attend the course will be:
(a) 2 or fewer
(b) exactly 4?

Q2 Vehicles approaching a T-junction must either turn right or turn left. It is observed that 43% turn right. Find the probability that of a random sample of eight vehicles, approaching the junction, the number turning right will be:
(a) exactly four
(b) fewer than two
(c) six or fewer.

Q3 A golfer practises on a driving range. He counts 'success' as driving a ball within 15 m of the flag. The probability of 'success' with each particular drive is 0.3.

If he drives ten balls, find the probability of
(a) four or fewer 'successes'
(b) exactly four 'successes'
(c) from two to five (inclusive) 'successes'
(d) four or fewer 'failures'.

Find the mean and standard deviation of the number of 'successes' in 20 drives.

Q4 Items from a production line are examined for any defects. The probability that any item will be found to be defective is 0.15, independently of all other items.
(i) A batch of 16 items is inspected. Using tables of cumulative binomial probabilities, or otherwise, find the probability that:
    (a) at least four items in a batch are defective
    (b) exactly four items in a batch are defective.
(ii) Five batches, each containing 16 items, are taken.
    (a) Find the probability that at most two of these five batches contain at least four defective items.
    (b) Find the expected number of batches that contain at least four defective items.

Q5 In the first round of a longjump competition each competitor makes four jumps. If the competitor oversteps a line the jump is disallowed. The numbers of jumps disallowed for the 40 competitors are summarised below.

Number of jumps disallowed	0	1	2	3	4
Number of competitors	14	9	5	6	6

(a) Calculate the mean and standard deviation of the number of jumps disallowed per competitor.

(b) Calculate the proportion p of jumps disallowed.

(c) Assuming the number of jumps disallowed for each competitor follows a binomial distribution, use your calculated value of p to estimate the mean and variance of this distribution.

(d) Do you think the binomial distribution is an adequate model for the data above? Give a reason.

(e) Give a reason, apart from any numerical calculations, why the number of jumps disallowed may not follow a binomial distribution.

Q6 A doctor wishes to undertake a trial into the effectiveness of a new treatment for a skin condition. She asks patients to take part in the trial and observes that the probability of them agreeing is 0.3 and may be assumed to be independent of whether or not other patients agree.

(a) If she asks 25 patients to take part in the trial, find the probability that:
   (i)   four or fewer will agree
   (ii) exactly 19 will **not** agree.

The probability that a patient who agrees to take part in the trial withdraws before the end is 0.23.

(b) For **each** of the following cases state giving a reason, whether the binomial distribution is likely to provide an adequate model for the random variable R.
   (i)   Out of 100 patients taking part in the trial, R is the total number withdrawing before the end.
   (ii) S is the total number of patients asked, in order to obtain 100 to take part in the trial.

Answers can be found on page 126.

20 Normal distribution

Key points to remember

- The normal distribution is a **continuous** distribution.
- Probability is represented by the **area under a curve**.

  (i) The total area under this curve is 1.
  (ii) The curve is bell-shaped.
  (iii) The curve is symmetrical.

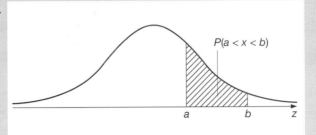

- Normal distribution tables list a **standard normal distribution**, that is a normal distribution with mean 0 and standard deviation 1. This is usually denoted z.

- To make calculations you must:

  ■ **either** standardise x, an observation from a normal distribution with mean μ and standard deviation σ using:

  $$z = \frac{x - \mu}{\sigma}$$

  ■ **or** convert standard normal values to x by rearranging the formula:

  $$x = \mu + z\sigma$$

- Tables enable you to find p for a given value of z, and z for a given value of p.

- **Remember** that the standard normal distribution has mean zero. Negative values of z are in the lower half of the distribution and positive values of z are in the upper half.

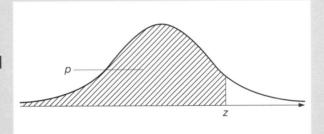

Formulae you must know

- $z = \dfrac{x - \mu}{\sigma}$

Don't make these mistakes ...

Never subtract x from the population mean μ, **always subtract μ from x.**

Don't try to work out in your head how to combine probabilities. Always draw a diagram.

Q1 Tins of peas are filled and sealed by a machine.

The weight of peas in a tin is normally distributed with mean 435 g and standard deviation 25 g. Find the probability that the weight of the peas in a randomly selected tin:

(a) is less than 470 g
(b) exceeds 465 g
(c) lies between 410 g and 430 g.

(a) $z = \dfrac{470 - 435}{25} = 1.4$

$P(X < 470) = 0.919$

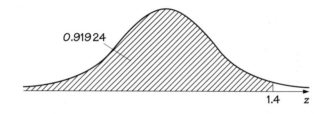

- Note $\mu = 435$, $\sigma = 25$

- Standardise 470 using $z = \dfrac{(x - \mu)}{\sigma}$.

- Draw a diagram and identify the area you need.

- The shaded area is given by your tables.

- The tables give five places of decimals, but it is sensible to round your final answer to three significant figures.

(b) $z = \dfrac{465 - 435}{25} = 1.2$

$P(X > 455) = 1 - 0.884\,93 = 0.115$

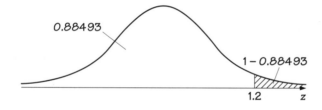

- Standardise 465.

- You can use the fact that the total area under the curve is 1 to find the required area.

(c) $z_1 = \dfrac{410 - 435}{25} = -1.0$

$z_2 = \dfrac{430 - 435}{25} = -0.2$

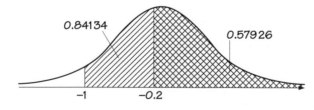

$P(410 < X < 430)$
$= 0.841\,34 - 0.579\,26 = 0.262$

- Standardise 410 and 430. Note that the negative signs mean that the z-values are in the lower half of the distribution.

- Now use symmetry. The area above -1 is the same as below 1.

- From the diagram, you should see that the area you need is the difference between the two areas you found from tables.

- Keep as many significant figures as possible in the calculation but round the final answer to three significant figures.

Q2 A health food cooperative sells free-range eggs.

Eggs weighing less than 48 g are graded small, those weighing more than 59 g are graded large. All other eggs are graded medium.

The weights of eggs from a particular supplier may be modelled by a normal distribution with mean 52 g and the standard deviation 4 g.

(a) Find the proportion of eggs graded medium.

It is decided to brand the largest 2% of eggs from this supplier as 'jumbo' eggs.

(b) Find the weight of the lightest egg which would be branded 'jumbo'.

(c) Find the median weight of the eggs graded 'small'.

- Note that $\mu = 52$ and $\sigma = 4$.

- The proportion is the same as the probability of a randomly-selected egg being graded medium.

(a) $z_1 = \dfrac{48 - 52}{4} = -1$

$z_2 = \dfrac{59 - 52}{4} = 1.75$

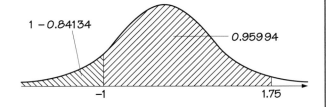

The proportion of eggs graded medium
$= 0.959\,94 - 0.158\,66 = 0.801$

- Standardise 48 and 59.
- Note the z-values have different signs.
- The diagram shows that the area you need is the area below 1.75 *minus* the area below -1. To find the area below -1 you need to use symmetry, remembering that the total area under the curve is 1.

- Keep as many figures as possible in the calculation but round the answer to three significant figures.

(b) The weight of the lightest 'jumbo' egg

$= 52 + 2.0537 \times 4$

$= 60.2\,g$

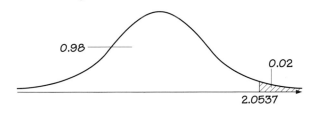

- Now you need to use your tables to find the z-value which is exceeded by exactly 2% or 0.02 of the population. That is the z-value which exceeds 0.98 of the population.

- Now convert the z-value into grams using $x = \mu + z\sigma$.

- Round to three significant figures.

(c) From part (a), the smallest 0.158 66 of the eggs are graded small. The median of these eggs will exceed half of 0.158 66 or 0.079 33 of the population.

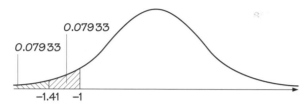

Median of eggs graded small is

$52 - 1.41 \times 4 = 46.4\,g$

- Use your tables to find the z-value exceeded by 0.920 67 of the population.
- Note this must be negative.
- You could increase accuracy by interpolation but it is much easier and sufficiently accurate to round the proportion to 0.92.
- Now convert the z-value into grams.

Q3 A group of students believes that the time taken to travel to college, T minutes, can be assumed to be normally distributed. Within the college 5% of students take at least 55 minutes to travel to college and 0.1% take less than 10 minutes.

Find the mean and standard deviation of T.

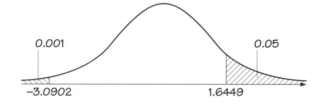

$\mu + 1.6449\sigma = 55$

$\mu - 3.0902\sigma = 10$

Solving these equations gives $\mu = 39.4$ minutes, $\sigma = 9.50$ minutes.

- Find the z-value exceeded by 99.9% or 0.999 of the population, and the z-value which exceeds 95% or 0.95 of the population.
- Take care with the signs.
- Use $x = \mu + z\sigma$ (x, z known; μ, σ unknown) to obtain two simultaneous equations. Then solve them.

Questions to try

Q1 The random variable X has a normal distribution with a mean of 36 and a standard deviation of 2.5. Find $P(X < 32)$.

Q2 The lengths of components produced by a machine are normally distributed with a mean of 0.984 cm and a standard deviation of 0.006 cm. The specification requires that a component should measure between 0.975 cm and 0.996 cm in length. Find the probability that a randomly-selected component will meet the specification.

Q3 A smoker's blood nicotine level, measured in ng/ml, may be modelled by a normal random variable with mean 310 and standard deviation 110.

(a) What proportion of smokers have blood nicotine levels lower than 250?
(b) What blood nicotine level is exceeded by 20% of smokers?

Q4 The operational lifetimes of certain electronic components are found to be normally distributed with mean 5200 hours and standard deviation 400 hours.

Calculate, to three significant figures:

(a) the probability that a component selected at random has a lifetime less than 5800 hours
(b) the proportion of such components having lifetimes between 4500 and 5800 hours
(c) the 84th percentile of this distribution.

Q5 Consultants employed by a large library reported that the time spent in the library by a user could be modelled by a normal distribution with mean 65 minutes and standard deviation 20 minutes.

(a) Assuming that this model is adequate, what is the probability that a user spends:
    (i) less than 90 minutes in the library
    (ii) between 60 and 90 minutes in the library?

The library closes at 9.00 p.m.

(b) Explain why the model above could not apply to a user who entered the library at 8.00 p.m.
(c) Estimate an approximate latest time of entry for which the model above could still be plausible.

Q6 Following extensive research, a clothes manufacturer concluded that adult male customers requiring T-shirts will have chest measurements which may be modelled by a normal distribution with mean 101 cm and standard deviation 5 cm. It is decided to make three sizes:

large to fit chests over 108 cm
medium to fit chests between 98 cm and 108 cm
small to fit chests of less than 98 cm.

(a) Assuming the model is adequate, calculate the proportion of adult male customers who will require:
    (i) **large** T-shirts
    (ii) **medium** T-shirts.

It is decided to introduce an **extra large** size to fit the largest 2% of chests of adult male customers.

(b) Find the minimum chest measurement for **extra large** T-shirts.
(c) Find the median chest measurement of those customers who require **small** T-shirts.

Answers can be found on page 127.

Key points to remember

- **Correlation** and **regression** are about **relationships** between two variables.

- **Simple linear regression** uses observed data to estimate an equation of the form $y = a + bx$ which relates a response (or dependent) variable, y, to an explanatory variable x. This is known as the **regression of y on x**.

- The variables are not interchangeable. For example, the depth of a river, y, may depend on the previous week's rainfall, x. The previous week's rainfall cannot depend on the depth of the river. Although it is possible to calculate the regression equation of rainfall on depth of river, it is meaningless. You should only calculate the regression equation of depth of river on rainfall.

- The equation may be used to estimate the value of y corresponding to a particular value of x, but only if the relationship is approximately linear and the value of x is within the range of the observed values.

- The **product moment correlation coefficient** measures the strength of a linear relationship between two variables. It is measured on a scale from -1 to $+1$.

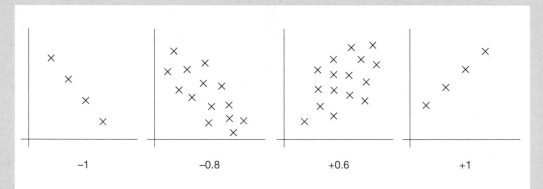

- In correlation the variables x and y are interchangeable.

- **Spearman's rank correlation** coefficient may be calculated by replacing the raw values of x and y by their ranks and then calculating the product moment correlation coefficient between ranks.

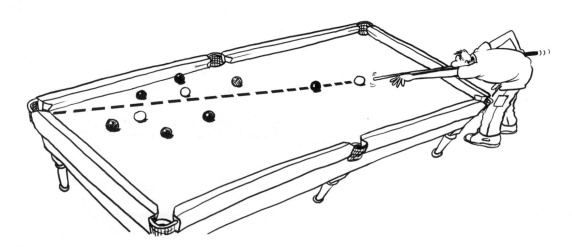

Formulae you must know

Provided you have a calculator with statistical functions there are no formulae you must know. The regression and correlation coefficients may be obtained directly from your calculator – at least in real life. If your examination board does not give you the raw data, but only summations, you may need some of these formulae.

Given n pairs of observed values (x, y):

for the regression equation $y = a + bx$

- $b = \dfrac{\Sigma(x - \bar{x})(y - \bar{y})}{}$

which may be calculated directly or by using the formula:

- $b = \dfrac{\Sigma xy - \dfrac{\Sigma x \Sigma y}{n}}{\Sigma x^2 - \dfrac{(\Sigma x)^2}{n}}$

- $a = \bar{y} - b\bar{x}$

The product moment correlation coefficient, r, is given by:

- $r = \dfrac{\Sigma(x - \bar{x})(y - \bar{y})}{\sqrt{\Sigma(x - \bar{x})^2 \Sigma(y - \bar{y})^2}}$

which may be calculated directly or by using the formula:

- $r = \dfrac{\Sigma xy - \dfrac{\Sigma x \Sigma y}{n}}{\sqrt{\left(\Sigma x^2 - \dfrac{(\Sigma x)^2}{n}\right)\left(\Sigma y^2 - \dfrac{(\Sigma y)^2}{n}\right)}}$

Spearman's rank correlation coefficient may be found by using ranks in the formula for r. An alternative formulation is:

- $r_s = 1 - \dfrac{6\Sigma d^2}{n(n^2 - 1)}$

where d is the difference between the x-rank and the y-rank for an observation. This gives the same result as the product moment correlation coefficient between ranks, provided there are no tied ranks. If there are tied ranks the result will be slightly different. In this case the correct formula to use is the product moment correlation coefficient between ranks. However you are unlikely to be penalised in an examination for using a different formula.

Don't make these mistakes...

Don't calculate the regression equation of p on q if the variable q depends on the variable p. Instead, calculate the regression equation of q on p.

Don't extrapolate – that is, don't use the regression equation of y on x to predict y for values of x outside the range of the observed values of x.

Don't calculate the product moment correlation coefficient for variables which are not approximately linearly related or which contain outliers. It may be appropriate to calculate Spearman's rank correlation coefficient in these circumstances.

Don't claim that changes in x cause changes in y if your only evidence is a high correlation coefficient between x and y. This shows that the variables are associated (perhaps through another variable) but not that changing one causes the other to change.

Q1 The following data were extracted from a daily newspaper for eight days in October.

Rainfall, x cm	1.3	3.8	4.2	2.6	2.1	2.6	5.3	0.9
Sunshine, y hours	1.5	0.3	0.0	4.2	3.6	0.5	0.0	1.4

(a) Calculate Spearman's rank correlation coefficient for the data.

(b) Comment on the value you have obtained.

(a)

Rank, x	7	3	2	4.5	6	4.5	1	8
Rank, y	3	6	7.5	1	2	5	7.5	4

From the calculator, $r_s = -0.711$

(b) The correlation coefficient shows some tendency for days with the most rainfall to have the least sunshine.

- The data have been ranked from largest to smallest. They could have been ranked from smallest to largest, but, in order to interpret the coefficient, both x and y should be ranked in the same direction.

- One rainfall of 2.6 cm should be ranked 4 and the other ranked 5. From the given data it is not possible to decide which is larger and so both have been given the mean of 4 and 5, i.e. 4.5.

- This is the product moment correlation coefficient between the ranks.

- The formula based on Σd^2 gives -0.690 in this case. The relatively large difference is due to the relatively large proportion (25%) of tied ranks.

- The negative sign shows high rainfall associated with low sunshine. The magnitude shows that the association is moderately strong.

Q2 **(a)** Estimate, without undertaking any calculations, the value of the product moment correlation coefficient in each of the scatter diagrams below.

(i)

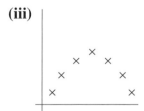

(ii)

(iii)

(iv)

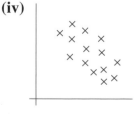

(b) Why is the product moment correlation coefficient an unsuitable measure to use for diagram (iii)?

(a) (i) *0.0*

   (ii) *0.95*

   (iii) *0.0*

   (iv) *−0.5*

(b) *A non-linear relationship is indicated.*

- If the data show an upward trend the coefficient will be positive, if they show a downward trend the coefficient will be negative.

- The closer to a straight line the points lie, the closer the magnitude of the coefficient is to 1.

- If you imagine moving the origin to $(\bar{x}, \bar{y})$ then points that lie in the first and third quadrants will make a positive contribution and points that lie in the second and fourth quadrants will make a negative contribution.

Q3 **(a)** Suppose you were asked to relate the variables r and s with a regression line of the form $y = a + bx$ where r is the population of a country and s is the number of medals won at the Sydney Olympic Games.

   (i) State, giving a reason, which of the variables you would choose for x and which for y.

   (ii) Would you expect b to be positive or negative?

(b) Repeat part (a) if:
   r is the temperature of a cup of tea and
   s is the time which has elapsed since it was poured out.

(a) (i) *$r \to x$ and $s \to y$ because the number of medals won may depend on the size of the population, but the size of the population cannot depend on the number of medals won.*

   (ii) *Positive – it is likely that countries with larger populations will win more medals.*

(b) (i) *$s \to x$ and $r \to y$ because the temperature will depend on how long the cup of tea has been poured out, the time cannot depend on the temperature of the cup of tea.*

   (ii) *Negative, as time passes the temperature will reduce.*

Q4 The table below shows x, a measure of social deprivation of an area (the higher the value of x, the more deprived the area) and y, the percentage of pupils at the local comprehensive school gaining four or more GCSE passes at grade C or above.

School	A	B	C	D	E	F	G	H	I	J	K
x	3.5	28.3	17.9	4.6	12.2	13.5	53.2	67.1	36.2	14.7	26.4
y	78.7	51.3	57.5	72.9	70.4	52.4	41.2	19.0	44.9	64.4	48.8

(a) Draw a scatter diagram of the data.

(b) Calculate the regression line of y and x and draw it on your diagram.

(c) The headteacher of school F claims that it is a better school than G. Comment on this claim.

(a)

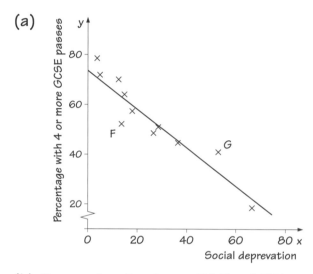

(b) Regression line is $y = 74.7 - 0.791x$

$x = 0$, $y = 74.7$ $x = 60$, $y = 27.2$

- You can find these values from the calculator. It is reasonable to use three significant figures. To draw the line, calculate the coordinates of two points at opposite ends of the graph and join them. You should calculate a third point as a check.

(c) The diagram clearly shows an association between social deprivation and percentage GCSE passes. School F has a higher percentage pass rate than School G. However, School F is below the regression line indicating a lower pass rate than would be expected from its social deprivation index. School G is above the regression line indicating a higher percentage pass rate than would be expected from its social deprivation index. Even if the percentage of GCSE passes is accepted as a valid measure of a school's worth, there is no basis for the headteacher's claim.

- If the line does not pass through the points plotted on the scatter diagram, you have made a mistake. Check your calculation of the regression line.

- Look for comments which relate to the data you have been given.

- It is OK to make general comments, such as questioning the value of the measures used, in addition to but not instead of making comments based on the data.

Q1 During the lambing season, eight ewes and the lambs they bore were weighed at the time of birth.

Ewe	A	B	C	D	E	F	G	H
Weight of ewe, kg	49	46	48	45	46	42	43	40
Weight of lamb, kg	3.7	3.0	3.4	2.9	3.1	2.7	3.0	2.8

Calculate the product moment correlation coefficient between the weights of the ewes and the weights of their lambs.

Comment, briefly, on your result.

Q2 As part of a practical exercise in statistics, Miriam was shown photographs of 11 people and asked to estimate their ages. The actual ages and the estimates made by Miriam are shown below.

Actual age, x	86	55	28	69	45	7	17	11	37	2	78
Miriam's estimate, y	88	60	35	77	50	8	15	6	49	2	85

(a) Draw a scatter diagram of Miriam's estimate, y, and the actual age, x.
(b) Calculate the equation of the regression line of Miriam's estimate on actual age.
(c) Draw this regression line on your scatter diagram. Draw also the line $y = x$.

Comment on Miriam's estimates.

Q3 The following data show the annual income per head, x, (in US$) and the infant mortality, y, (per thousand live births) for a sample of 11 countries.

Country	A	B	C	D	E	F	G	H	I	J	K
x	130	5950	560	2010	1870	170	390	580	820	6620	3800
y	150	43	121	53	41	169	143	59	75	20	39

(a) Draw a scatter diagram of the data. Describe the relationship between income per head and infant mortality suggested by the diagram.
(b) An economist asks you to calculate the product-moment correlation coefficient.
   (i) Carry out this calculation.
   (ii) Explain briefly to the economist why this calculation may not be appropriate.
(c) Calculate Spearman's rank correlation coefficient between x and y.
(d) Comment on and compare the values of the two correlation coefficients you have calculated.

Q4 A drilling machine can run at various speeds, but in general the higher the speed the sooner the drill needs to be replaced. Over several months, 15 pairs of observations relating to speed, s revolutions per minute, and life of drill, h hours, are collected.

For convenience the data are coded so that $x = s - 20$ and $y = h - 100$ and the following summations obtained.

$\Sigma x = 143$; $\Sigma y = 391$; $\Sigma x^2 = 2413$; $\Sigma y^2 = 22\,441$; $\Sigma xy = 484$

(a) Find the equation of the regression line of h on s.
(b) Interpret the slope of your regression line.
(c) Estimate the life of a drill revolving at 30 revolutions per minute.

Q5 **(a)** A road haulage contractor owns four lorries of the same age and specification. She employs four drivers: Ahmed (A), Beryl (B), Chris (C) and Danny (D). She collects data, for a number of long journeys, on the driver, load carried and the diesel consumption.

Driver	A	B	D	A	D	C	C	A	D	B
Load x (kg)	5650	10 100	7800	8450	5500	6950	7600	8300	6250	6600
Diesel consumption y (km/litre)	6.22	5.18	5.25	5.49	6.01	5.99	5.89	5.42	5.77	6.11

(i) Draw a scatter diagram of y against x.
(ii) Calculate the equation of the regression line of y on x and draw it on your scatter diagram.
(iii) Give an interpretation of the slope and of the intercept of the regression line.
(iv) Why would it be unwise to use the regression equation to predict the diesel consumption if the load was 30 000 kg?
(v) Comment on the diesel consumption of lorries driven by Danny.
(vi) Why was the regression line of y on x calculated rather than the regression line of x on y?

Answers can be found on pages 127–128.

1 Algebra and equations 1

Q1

(a) $a^5 \times a^{-2} = a^{5-2} = a^3$

(b) $\dfrac{a^5}{a^{-3}} = a^{5-(-3)} = a^8$

(c) $\left(\dfrac{\sqrt[4]{a}}{\sqrt[7]{a}}\right)^3 = \left(a^{\frac{1}{4}-\frac{1}{7}}\right)^3$

$= \left(a^{\frac{3}{28}}\right)^3$

$= a^{\frac{9}{28}}$

How to solve:
$a^m \times a^n = a^{m+n}$

$(a^m)^n = a^{mn}$

Q2

(i) $k = -4\sqrt{3}$

$\Rightarrow f(x) = x^2 - 4\sqrt{3}x + 9$

$\Rightarrow f(x) = (x - 2\sqrt{3})^2 - 3$

$a = -2\sqrt{3}$ and $b = -3$.

The least value of f(x) is −3 (occurs when $x = 2\sqrt{3}$).

How to solve: For two distinct real roots:
$k^2 - 36 > 0$
$\Rightarrow k < -6$ or $k > 6$ i.e. $|k| > 6$.

First, substitute for the given value of k. Then to complete the square, the value of a is half the coefficient of x.

(ii) $f(x) = 0$

$\Rightarrow (x - 2\sqrt{3})^2 - 3 = 0$

$x = 2\sqrt{3} \pm \sqrt{3}$

i.e. $x = 3\sqrt{3}$ or $\sqrt{3}$.

How to solve: You do not need to use the formula here since (i) gives the answer easily.

Q3 $x^2 > x + 20$

$\Rightarrow x^2 - x - 20 > 0$

$\Rightarrow (x - 5)(x + 4) > 0$

$\Rightarrow x < -4$ or $x > 5$

How to solve: You need to factorise and remember that f(x) > 0 if the factors are both positive or both negative.

Q4 $x + x + 1 > 10$

$2x + 1 > 10$

$x > 4.5$

$x(x + 1) < 72$

$x^2 + x - 72 < 0$

$(x - 8)(x + 9) < 0$

$-9 < x < 8$

So the numbers can be 5 and 6, 6 and 7 or 7 and 8.

How to solve: Let the numbers be x and $x + 1$. Form and solve an equality by considering the sum of the two numbers.

Form and solve an equality by considering the product of the two numbers.

Remember that you are looking for pairs of consecutive numbers.

Q5

How to solve: You should always draw a diagram and mark in the side lengths.

(a) $x + x - 5 + x + x - 5 > 32$

$\Rightarrow 4x - 10 > 32$

$\Rightarrow 4x > 42$

$\Rightarrow x > 10.5$

(b) $x(x - 5) < 104$

(c) $x^2 - 5x - 104 < 0$ from (b)

$\Rightarrow (x - 13)(x + 8) < 0$

$\Rightarrow -8 < x < 13$

since $x > 10.5$ from (a), then $10.5 < x < 13$

Q6

(i) $x^2 - 3x + 2 = 3x - 7$

$x^2 - 6x + 9 = 0$

$(x - 3)^2 = 0 \Rightarrow x = 3$

(ii) $y = 3x - 7$ is a tangent to $y = x^2 - 3x + 2$.

How to solve: Form a quadratic by equating the two expressions for y.

Since there is only one solution this means that the line touches the parabola once, i.e. it is a tangent.

2 Arithmetic and geometric progressions

Q1 $a = 7$, $d = 2$

$u_{18} = 7 + (18 - 1) \times 2 = 41$

$S_{15} = \dfrac{1}{2} \times 15 \times (2 \times 7 + (15 - 1) \times 2)$

$= 315$

How to solve: First note that this is an AP. Identify the first term and the common difference.

Substitute $n = 18$, $a = 7$ and $d = 2$ into the formula for the nth term of an AP.

Substitute $n = 15$, $a = 7$ and $d = 2$ into the formula for the sum of n terms of an AP.

Q2 $a = 100$, $r = 0.9$

(a) $u_{10} = 100 \times 0.9^9$

$= 38.742$

(b) $S_{20} = 100 \left(\dfrac{1 - 0.9^{20}}{1 - 0.9}\right)$

$= 878.4$ (1 d.p.)

(c) $S_\infty = \dfrac{100}{1 - 0.9}$

$= 1000$

How to solve: You should notice that this is a GP with first term 100 and common ratio 0.9.

Use the formula for the nth term of a GP, with $n = 10$.

Use the formula for the sum of n terms of a GP with $n = 20$.

As $|r| < 1$, you can use the formula for the sum to infinity.

Answers

Q3

(a) $S_{20} = \frac{1}{2} \times 20(2 \times 4 + (20 - 1) \times 3))$

$= 650$

(b) $4 + 3(n - 1) > 103$

$1 + 3n > 103$

$n > 34$

Q4 $a + 7d = 40$

$a + 19d = 124$

$\therefore d = 7$ and $a = -9$

$S_{20} = \dfrac{20(2 \times -9) + (20 - 1 \times 7)}{2}$

$= 1150$

Q5

(a) $ar = 80$

$ar^4 = 5.12$

$\therefore r = \dfrac{2}{5}$ and $a = 200$

(b) $S_\infty = \dfrac{200}{1 - \frac{2}{5}} = \dfrac{1000}{3}$

(c) $S_{14} = 200\left(\dfrac{1 - \left(\frac{2}{5}\right)^{14}}{1 - \left(\frac{2}{5}\right)}\right)$

$= 333.332\,44$

$S_\infty - S_{14} = 8.9 \times 10^{-4}$

Q6 $ar = 1$

$\dfrac{a}{1 - r} = 7.2$

$7.2r^2 - 7.2r + 1 = 0$

$\therefore r = \dfrac{1}{6}$ or $r = \dfrac{5}{6}$

For $r = \dfrac{5}{6}$, $a = \dfrac{6}{5}$ and the

sequence begins

$\dfrac{6}{5}, 1, \dfrac{5}{6}, \dfrac{25}{36}, \cdots$

How to solve these questions

Q3

(a) Use the formula for the sum of n terms of an AP with $n = 20$, $a = 4$ and $d = 3$.

(b) Form and solve an inequality based on the formula for the nth term of an AP, in this case $u_n > 103$.

Q4 By using the formula for the nth term of an AP you can form a pair of simultaneous equations.

Solve them to find d and a.

Once you know d and a you can use the formula for the sum of n terms with $n = 20$ to find the required sum.

Q5 First, use the formula for the nth term of a GP to form a pair of simultaneous equations.

Solve them to find r and a.

As $|r| < 1$, you can use the formula for the sum to infinity.

First use the formula for the sum of the first n terms of a GP to find the sum of the first 14 terms.

Then find the difference between the two sums.

Q6 You can form this equation because the second term is 1.

You can form this equation as the sum to infinity is 7.2.

Substitute $a = \dfrac{1}{r}$ in the second equation and simplify to find this quadratic equation.

Then you can solve the quadratic to obtain these solutions.

Use the larger value of r to find a.

Answers

$S_{20} = \dfrac{6}{5}\left(\dfrac{1 - \left(\frac{5}{6}\right)^{20}}{1 - \left(\frac{5}{6}\right)}\right)$

$= 7.012$ (3 d.p.)

Q7 $1932 = 1200 \times \left(1 + \dfrac{r}{100}\right)^4$

$1 + \dfrac{r}{100} = \sqrt[4]{\dfrac{1932}{1200}}$

$r = 12.64\%$ (correct to 2 d.p.)

$S_{10} = 1200\left(\dfrac{1 - \left(1 + \frac{r}{100}\right)^{10}}{1 - \left(1 + \frac{r}{100}\right)}\right)$

$= £21\,721$

Q8 A: $S_{10} = \dfrac{1}{2} \times 10 \times$

$(2 \times 1000 + (10 - 1) \times 100)$

$= £14500$

B: $14\,500 = x\left(\dfrac{1 - 1.1^{10}}{1 - 1.1}\right)$

$x = £909.81$

Q9 $d_2 = 2 + \dfrac{2}{5} \times 2 + \dfrac{2}{5} \times 2$

$= 3.6\,m$

$d_\infty = 2 + \dfrac{2}{5} \times 2 + \dfrac{2}{5} \times 2 + \dfrac{2}{5} \times 2 + \left(\dfrac{2}{5}\right)^2$

$\times 2 + \left(\dfrac{2}{5}\right)^2 \times 2 + \ldots$

$= 2 + 2 \times \dfrac{\left(2 \times \frac{2}{5}\right)}{\left(1 - \frac{2}{5}\right)}$

$= 2 + 2 \times \dfrac{4}{3}$

$= \dfrac{14}{3}$

How to solve these questions

Then use the formula for the sum of n terms of a GP.

You can use the fact that the 5th term is 1932 to form this equation.

Solve the equation to find r.

Then use the formula for the sum of n terms to find the total that is paid into the fund.

Use the formula for the sum of an AP to find the total for scheme A.

Form and solve an equation based on a GP with 10 terms and sum 14 500, using the formula for the sum of n terms of a GP.

The total distance is made up of a fall of 2 m, a rise of $\dfrac{2}{5} \times 2$ and a fall of the same distance.

The total distance is given by this expression which is 2 plus twice a GP with first term $\dfrac{2}{5} \times 2$ and common ratio $\dfrac{2}{5}$. You can sum the GPs using the formula for the sum to infinity.

Answers

Q10

(a)

$\frac{1}{4}$ of nth term $= \frac{1}{4} \times 8 \times 0.2^{n-1}$

$= 2 \times 0.2^{n-1}$

$= 10 \times 0.2^n$

$S_\infty = \frac{8}{1 - 0.2} = 10$

Difference $= 10 - 8 \times \left(\frac{1 - 0.2^n}{1 - 0.2} \right)$

$= \mathbf{10 \times 0.2^n}$

(b) $\ln(ab^n) - \ln(ab^{n-1}) = \ln b$

As $\ln b$ is constant this is an AP:

$\frac{1}{2} n(2\ln 2 + (n-1)\ln 3)$

$= \frac{1}{2} n(2\ln 3 + (n-1)\ln 2)$

$(n-3)\ln 3 = (n-3)\ln 2$

$\mathbf{n = 3}$

---

3 TRIGONOMETRY

Q1

$r\theta = 2r$

$\theta = 2$

Area $= \frac{1}{2} \times r^2 \times 2 = \mathbf{r^2}$

Q2

Area of sector

$= \frac{1}{2} \times 6^2 \times 0.7 = 12.6\,\text{cm}^2$

Area of triangle

$= \frac{1}{2} \times 6 \times 6 \sin 0.7$

$= 11.60\,\text{cm}^2$

Area shaded $= 12.60 - 11.60$

$= \mathbf{1.00\,cm^2}$

Perimeter

$= 6 \times 0.7 + 2 \times 6 \sin 0.35$

$= \mathbf{8.31\,cm}$

How to solve these questions

You need to find an expression for $\frac{1}{4}$ of the nth term.

You also need to find the sum to infinity.

Then taking the difference between the sum to infinity and the sum of n terms gives you the required result.

You need to find the difference between consecutive terms. As this is constant the series is an AP.

You should form an equation by equating the sum of n terms for each series. Then you can solve the equation to find n.

Use the formula for arc length to find θ.

Then you can use this value of θ to find the area.

You should first find the area of the sector.

When you find the area of the triangle the height is given by $6 \sin 0.7$.

Answers

Q3

(a) $20 = 2r + r\theta$

$\theta = \frac{20 - 2r}{r}$

Area $= \frac{1}{2} \times r^2 \times \left(\frac{20 - 2r}{r} \right)$

$= 10r - r^2$

$= \mathbf{25 - (r-5)^2}$

(b) **Maximum $A = 25$ when $r = 5$.**

Q4

$\sin^{-1}(-0.5) = -30°$

$\theta = 180° + 30° = 210°$ or

$\theta = \mathbf{360° - 30° = 330°}$

Q5

$\sin 2x = \frac{1}{2}$

$2x = 30°, 150°, 390°$ or $510°$

$\mathbf{x = 15°, 75°, 195°}$ or $\mathbf{255°}$

Q6

$3 \dfrac{\sin\theta}{\cos\theta} = \dfrac{1}{\cos\theta}$

$3\sin\theta = 1$

$\sin\theta = \dfrac{1}{3}$

$\theta = \mathbf{0.340}$ or

$\theta = \pi - \mathbf{0.340 = 2.802}$

Q7

$2(1 - \cos^2 x) + 5\cos x = 0$

$2\cos^2 x - 5\cos x - 2 = 0$

$\cos x = 2.851$ or -0.3508

$\cos^{-1} -0.3508 = \mathbf{110.5°}$

or $x = \mathbf{360° - 110.5°}$

$= \mathbf{249.5°}$

Q8

$15(1 - \sin^2\theta) = 13 + \sin\theta$

$15\sin^2\theta + \sin\theta - 2 = 0$

$\sin\theta = -\dfrac{2}{5}$ or $\dfrac{1}{3}$

$\theta = \mathbf{19.5°, 160.5°, 203.6°}$
or $\mathbf{336.4°}$

How to solve these questions

When finding the perimeter, note that it consists of two radii and the arc.

You should first consider the perimeter and obtain an expression for θ in terms of r.

Then you can substitute this into the formula for the area, and simplify.

You should note that A will be a maximum when the value of the term inside the bracket is 0.

You can obtain a first solution from your calculator. You can then calculate the other solutions, using a sketch graph if necessary.

You should first find the values of $2x$ that satisfy the equation.

The first of these is $\sin^{-1}\frac{1}{2} = 30°$.

Then you can calculate the others, but you must include solutions in the range $0 \le 2x \le 720°$. Finally, divide these solutions by 2 to obtain the values of x.

First, use $\tan\theta = \dfrac{\sin\theta}{\cos\theta}$

You can then find the solutions to this equation.

First, replace $\sin^2 x$ by $1 - \cos^2 x$.

This will give you a quadratic equation in $\cos x$.

You can solve this with the formula to give these two solutions. You find the first solution from your calculator. Subtracting this from 360° will give the second solution.

First, replace $\cos^2\theta$ by $1 - \sin^2\theta$.

This gives you a quadratic equation in $\sin\theta$.

You find two solutions when you solve this.

You can then find the values of θ.

Answers

Q9 $4\sin^2\theta - 2\sin\theta = 4(1-\sin^2\theta) - 1$

$8\sin^2\theta - 2\sin\theta - 3 = 0$

$\sin\theta = \frac{3}{4}$ or $-\frac{1}{2}$

This gives two exact values $\theta = \mathbf{210°}$ **and** $\theta = \mathbf{330°}$, and two which, correct to 1 d.p. are $\theta = \mathbf{48.6°}$ **and** $\theta = \mathbf{131.4°}$.

Q10

(a) (i) $10 - 3 = \mathbf{7\ m}$

(ii) $10 + 3 = \mathbf{13\ m}$

(b)

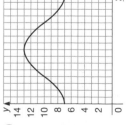

(c) $\frac{1}{2} \times \frac{2\pi}{k} = 6.2$

$k = \frac{\pi}{6.2}$

$= \mathbf{0.51}$ (2 d.p.)

4 COORDINATE GEOMETRY

Q1 Gradient of AB = $\left(\frac{14-6}{5-1}\right) = 2$

Equation of p is $y = 2x + c$.

$x = 1$ and $y = 6$ gives $c = 4$

Equation of p is $y = \mathbf{2x + 4}$.

The midpoint has coordinates $\left(\frac{1+5}{2}, \frac{6+14}{2}\right)$

$= (3, 10)$

Equation of q is $y = -\frac{1}{2}x + c$

$x = 3$ and $y = 10$ gives $c = \frac{23}{2}$.

How to solve these questions

First, replace $\cos^2\theta$ by $1 - \sin^2\theta$.

Then solve the quadratic in $\sin\theta$ to find two solutions.

You get the exact solutions from $\sin\theta = -\frac{1}{2}$.

You get these from $\sin\theta = \frac{3}{4}$.

For this you take $\cos kt = 1$.

For this you take $\cos kt = -1$.

You should show the minimum and maximum values, and that the period is $\frac{2\pi}{k}$.

You should note that half of a period is 6.2 hours.

You can then form and solve an equation for k.

First you need to find the gradient of the line through the two points.

You can then substitute the gradient into the equation of a straight line.

You can use the coordinates of A to find the value of c.

Then substitute the value of c in the equation.

The next step is to find the coordinates of the midpoint of AB.

As the lines are perpendicular, the gradient of $q = \frac{-1}{\text{gradient of } p} = -\frac{1}{2}$ and so you can substitute this value into the equation of a straight line.

You can the use the coordinates of the midpoint to find the value of c.

Answers

Q2

(a) $3(3x - 6) + x - 12 = 0$

$10x = 30$

$x = 3$

The coordinates are (3, 3).

(b) **Gradients are 3 and $-\frac{1}{3}$, so their product is −1, hence the lines are perpendicular.**

(c) The corners of the triangle are (3, 3), (2, 0) and (12, 0).

Area $= \frac{1}{2} \times (12 - 2) \times 3$

$= \mathbf{15}$ **square units**

Q3 The equation of p is

$y = \frac{1}{3}x + c$.

$x = -2$ and $y = 0$ gives $c = \frac{2}{3}$.

$\mathbf{y = \frac{1}{3}x + \frac{2}{3}}$ **or** $\mathbf{3y - x - 2 = 0}$

The equation of q is

$y = -3x + c$.

$x = 8$ and $y = 0$ gives $c = 24$.

$\mathbf{y = -3x + 24}$ **or** $\mathbf{y + 3x - 24 = 0}$

Q4

(a) The gradient of AC is

$\left(\frac{2-6}{7-3}\right) = -1$.

The equation is $y = -x + c$.

$x = 3$ and $y = 6$ gives $c = 9$.

$\mathbf{y = -x + 9}$ **or** $\mathbf{y + x - 9 = 0}$

The gradient of q is 1, hence the equation is $y = x + c$.

$x = 6$ and $y = 5$ gives $c = -1$.

$\mathbf{y = x - 1}$ **or** $\mathbf{y - x + 1 = 0}$

How to solve these questions

From the first equation $y = 3x - 6$.

Substitute this into the second equation and solve for x.

Substitute $x = 3$ into either equation to find the y-coordinate.

Find the gradients by writing the equations in the form $y = mx + c$ and then find their product.

As the lines intersect the x-axis you can find the coordinates of the other corners by substituting $y = 0$ into the equations.

You can now find the area. A sketch of the triangle may help you.

First, note that the gradient of the line will be 3 and substitute this in the equation of a straight line.

You can use the coordinates of the point that has been given to find c.

As the line q is perpendicular to p, you can deduce that its gradient will be 3.

The coordinates of the point on the line can be used to find c.

Find the coordinates of the point of intersection by equating the two equations and solving for x.

Substitute $x = 7$ into either equation to find the y-coordinate.

First, use the coordinates to calculate the gradient of the line AC.

Then write down the equation.

Now use one set of coordinates to find c and complete the equation.

Find the gradient, by using the fact that the product of the two gradients is −1.

This allows you to write down the equation.

The lines intersect when $-3x + 24 = \frac{1}{3}x + \frac{2}{3}$ which gives $x = 7$, so the coordinates are **(7, 3)**.

(b) The lines intersect when
$x - 1 = -x + 9$ or $x = 5$

The coordinates are **(5, 4)**.

The lines will intersect when they have the same y-coordinate.
You can use this to form and solve an equation for x.

Then substitute this value of x to find y.

Q5

(a) The gradient of AB
$$= \frac{-2-4}{6-(-2)} = -\frac{3}{4}.$$

First, find the gradient of the line AB.

This will allow you to write down the equation of the line.

The equation is $y = -\frac{3}{4}x + c.$

$x = -2$ and $y = 4$ gives $c = \frac{5}{2}.$

$y = -\frac{3}{4}x + \frac{5}{2}$ or $3x + 4y - 10 = 0$

Use the coordinates of A to find c and complete the equation of the line.

(b) The coordinates of D are
$$\left(\frac{-2+6}{2}, \frac{4+(-2)}{2}\right) = (2, 1)$$

First find the coordinates of the midpoint.

The gradient of CD
$$= \frac{5-1}{5-2} = \frac{4}{3}$$

Then you can calculate the gradient of CD.

Product of gradients
$$= -\frac{3}{4} \times \frac{4}{3} = -1, \therefore \text{ the lines}$$
are perpendicular.

You must show that the product of the gradients is –1.

5 FUNCTIONS

Q1

(a) $g(x) \geqslant 0$

Notice that the minimum value of $g(x)$ is 0.

(b) $gf(4) = g(4 - 3) = g(1)$
$= 4 \times 1^2 = 4$

First, find f(4) and then find g(1).

$fg(1) = f(4 \times 1^2) = f(4) = 1$

First, find g(1) and then find f(4).

(c) $gf(x) = g(x-3) = 4(x-3)^2$
Range is $\mathbf{gf(x)} \geqslant \mathbf{0}$

Apply f first and then g.
You should note that the minimum value of gf(x) is 0.

$fg(x) = f(4x^2) = \mathbf{4x^2 - 3}$
Range is $fg(x) \geqslant -3$

Apply g first and then f.
Note that the minimum value of fg(x) is –3.

(d) $y = x - 3$
$x = y + 3$
$\mathbf{f^{-1}: x \to x + 3}\ x \in \mathbb{R}$

Write $y = f(x)$ and solve for x.

Then write it as a function, using x instead of y.

Q2

(a) $\mathbf{f(x) > 0}$

You should note that $e^x > 0$ for all values of x.

(b) $gf(x) = g(e^x) = \mathbf{e^x + 1}$
Range is $\mathbf{gf(x) > 1}$

$fg(x) = f(x+1) = e^{x+1}$
Range is $\mathbf{fg(x) > 0}$

Apply f first and then g.
As $e^x > 0$, you should note that $e^x + 1 > 1$.
Apply g first and then f.
Remember that $e^{x+1} > 0$.

(c) $y = e^x$
$x = \ln y$
$\mathbf{f^{-1}: x \to \ln x}$
Domain $x > 0$
$y = x + 1$
$x = y - 1$
$\mathbf{g^{-1}: x \to x - 1}$
Domain $x \in \mathbb{R}$

Write $y = f(x)$ and solve for x. Then write as a function, using x instead of y.

You can use the same approach for g as you used for f.

Q3

(a) $y = \dfrac{x+6}{x}$

Write $y = f(x)$ and solve for x.

$xy = x + 6$
$x(y-1) = 6$
$x = \dfrac{6}{y-1}$

$\mathbf{f^{-1}: x \to \dfrac{6}{x-1}, x \in \mathbb{R}, x \neq 1}$

Then write as a function, using x instead of y.

(b) $\dfrac{x+6}{x} = \dfrac{6}{x-1}$

You need to form and solve the equation $f(x) = f^{-1}(x)$.

$(x+6)(x-1) = 6x$
$x^2 - x - 6 = 0$
$(x-3)(x+2) = 0$
$\mathbf{x = 3 \text{ or } x = -2}$

This produces a quadratic equation, so you must find two solutions.

Q4

(a)

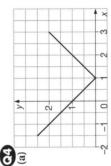

When sketching the graph you should show all the intersections of the curve with the axes.

There are two equations that you must solve.

$x - 1 = 7$ or $1 - x = 7$
$\mathbf{x = 8} \qquad \mathbf{x = -6}$

The first is when $x - 1$ is positive and the second is when $x - 1$ is negative.

(b) $gf(x) = g(x-1)$
$= (x-1)^2 - 1 = \mathbf{x^2 - 2x}$

Apply f first and then g.

Answers

(c)

x² - 2x = 1
x² - 2x - 1 = 0
$$x = \frac{2 \pm \sqrt{8}}{2} = 1 \pm \sqrt{2}$$

or

2x - x² = 1
x² - 2x + 1 = 0
(x - 1)² = 0
x = 1

Q5

(a) fg(x) = f(a − x) = **(a − x)²**
$$gf\left(\frac{\sqrt{a}}{2}\right) = g\left(\frac{a}{4}\right) = a - \frac{a}{4} = \frac{3a}{4}$$

(b)

a − x = 2   or   x − a = 2
x = a − 2   x = a + 2

Q6

(a) fg(x) = f(eˣ) = **eˣ + 2**
gf(x) = g(x + 2) = **e^{x+2}**

(b) **h⁻¹(x) = ln x − 2**
g⁻¹(x) = ln x
f⁻¹(x) = x − 2
f⁻¹g⁻¹(x) = f⁻¹(ln x)
= **ln x − 2 = h⁻¹(x)**

So the result is true

How to solve these questions

There are two equations that you must solve, to include the cases that gf(x) is positive or negative.

You should apply g first and then f.

Note that the graph intersects the vertical axis at *a* and the horizontal axis at *a*.

There are two equations that you must solve.

The first is when g(x) is positive and the second is when g(x) is negative.

Apply g first and then f.

Apply f first and then g.

Find h⁻¹(x) first, then find f⁻¹ and g⁻¹, so that you can find the composite function.

Apply g⁻¹ first, then f⁻¹ and compare the result with h⁻¹(x).

Answers

Q7

(a) Range is **f(x) > 0**

(b) y = ln(x + 1)
eʸ = x + 1
x = eʸ − 1
f⁻¹(x) = eˣ − 1

Q8

(a) $\dfrac{5}{x - 2} = x$
x² − 2x − 5 = 0
x = 1 ± √6

(b) Range is **g(x) ≥ 3**

(c) fg(x) = f(x² + 3)
$$= \frac{5}{x^2 + 3 - 2}$$
$$= \frac{5}{x^2 + 1}$$

(d) Range is **0 < fg(x) ≤ 5**
No not a one-to-one mapping

Q9

(a) **f⁻¹(x) ≥ 1**

(b) y = 1 + √x
x = (y − 1)²
f⁻¹(x) = (x − 1)²

(c) fg(x) = f(x²) = 1 + √(x²)
= 1 + x provided x ≥ 0

(d) fg(−2) = f(4)
= 1 + √4 = **3**

(e) fg(x) = 1 + √(x²)
= 1 + |x|

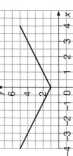

How to solve these questions

Note that if x > 0, then f(x) > ln 1.

Write y = f(x) and solve for x.

Write f as a function of x.

You must solve the equation f(x) = x.

Leave the square roots in your answer as it must be exact.

Apply g first and then f.

Note that the minimum value of g(x) is 3.

Note that the minimum value of the denominator of this fraction is 1 and also that it is always positive.

You should recognise that this is a many-to-one mapping and so no inverse will exist.

The domain of f⁻¹ will be the range of f, which is f(x) ≥ 1.

You should write y = f(x) and solve for x.

Write the function in terms of x.

Apply g and then f.

You should apply g and then f.

You need to show clearly the intersection with the vertical axis.

Q10

(a) $y = \ln(4 - 2x)$
$e^y = 4 - 2x$
$x = \dfrac{4 - e^y}{2}$
$f^{-1}(x) = \dfrac{4 - e^x}{2}$

(b)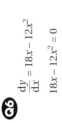

(c) $1.5 = 3^x$
$\ln 1.5 = x\ln 3$
$x = \dfrac{\ln 1.5}{\ln 3} = 0.369$

(d) $gf(1) = g(\ln 2)$
$= 3^{\ln 2} = 2.141$

6 DIFFERENTIATION I

Q1 $\dfrac{dy}{dx} = 4x^3 - 6x - 1$

Q2 $\dfrac{dy}{dx} = 3x^2 + 6$

Q3 $f'(x) = \dfrac{3}{2}x^{\frac{1}{2}} - x^{-\frac{3}{2}}$

Q4 $f'(x) = 2x - \dfrac{16}{x^2}$
$2x - \dfrac{16}{x^2} > 0$
$x^3 > 8$
$x > 2$

How to solve these questions

Write $y = f(x)$ and solve for x.

You should write the function in terms of x.

The curve meets the vertical axis when $x = 0$, and the horizontal axis when $y = 0$. You can use this to calculate the coordinates.

You should solve the equation $g(x) = 1.5$, by taking the logarithm of each side of the equation.

You should apply f and then g.

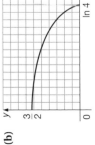

To differentiate $f(x)$, you first need to write each term in the form x^n. In this case you should find
$f(x) = x\sqrt{x} + \dfrac{2\sqrt{x}}{x} = x^{\frac{3}{2}} + 2x^{-\frac{1}{2}}$ which you can then differentiate.

Differentiate $f(x)$ to find $f'(x)$.

For an increasing function you will have $f'(x) > 0$, so you need to form and then solve this inequality.

How to solve these questions

First you need to expand the brackets, and then you differentiate each term.

When the gradient is 6, $\dfrac{dy}{dx} = 6$, so you need to form and solve this equation.

You can solve the quadratic by factorising.

You can find the coordinates by substituting the x-values into the expression for y.

You first need to differentiate to find $\dfrac{dy}{dx}$.

To find the stationary points you need to form and solve the equation obtained when $\dfrac{dy}{dx} = 0$.

You should think about how the gradients change at each stationary point. It will help if you draw up a table like this.

To find the set of values for which a function is decreasing you first need to find $\dfrac{dy}{dx}$ and then form and solve the inequality $\dfrac{dy}{dx} < 0$.

To find the area you should note that
$y = 20 - 2x$.
To find the maximum area, you first need to find $\dfrac{dA}{dx}$ and then solve the equation $\dfrac{dA}{dx} = 0$. Then you need to check that this gives a local maximum.

The gradient changes from positive to negative, so this is a local maximum.

To find the area you need to find expressions for the length and the width.

You can then multiply these together.

Answers

Q5

(a) $y = 2x^3 - x^2 + 6x - 3$
$\dfrac{dy}{dx} = 6x^2 - 2x + 6$

(b) $6 = 6x^2 - 2x + 6$
$0 = x(3x - 1)$
$x = 0$ or $x = \dfrac{1}{3}$

$(0, -3)$ and $\left(\dfrac{1}{3}, -\dfrac{28}{27}\right)$

Q6

$\dfrac{dy}{dx} = 18x - 12x^2$

$18x - 12x^2 = 0$
$x = 0$ or $x = 1.5$

x	-1	0	1	1.5	2
$\dfrac{dy}{dx}$	-30	0	6	0	-12

Local minimum at (0, 0) and local maximum at (1.5, 6.75)

$18x - 12x^2 < 0$
$6x(3 - 2x) < 0$
$x < 0$ or $x > \dfrac{3}{2}$

Q7

$A = xy$
$= x(20 - 2x)$
$= 20x - 2x^2$
$20 - 4x = 0$
$x = 5$
At $x = 4$, $\dfrac{dA}{dx} = 4$ and
at $x = 6$, $\dfrac{dA}{dx} = -4$

Maximum at $x = 5$

Q8

(a) Length $= 2x + 5$
Width $= \dfrac{63 - 2x - (2x + 5)}{2}$
$= 29 - 2x$
$A = (2x + 5)(29 - 2x)$

7 INTEGRATION I

Q1 $\int_0^3 (2x^3 - x^2)\,dx = \left[\dfrac{x^4}{2} - \dfrac{x^3}{3}\right]_0^3$

$$= \left(\dfrac{3^4}{2} - \dfrac{3^3}{3}\right) - 0$$

$$= \textbf{31.5}$$

You should simplify the expression so that it consists of terms of the form x^n. Include a constant of integration.

Q2 $\int \dfrac{x^2 - x}{\sqrt{x}}\,dx = \int (x^{\frac{3}{2}} - x^{\frac{1}{2}})\,dx$

$$= \dfrac{2}{5}x^{\frac{5}{2}} - \dfrac{2}{3}x^{\frac{3}{2}} + \boldsymbol{c}$$

You should simplify the expression so that it consists of terms of the form x^n. Include a constant of integration.

Q3 $\int \dfrac{1}{x^2}(\sqrt{x} - x)\,dx = \int (x^{-\frac{3}{2}} - x^{-1})\,dx$

$$= \textbf{-2}x^{-\frac{1}{2}} - \textbf{ln}|x| + \boldsymbol{c}$$

Q4 $x^2 = 12 - x$

$x^2 + x - 12 = 0$

$x = 3$ or $x = -4$

Area $= \int_0^3 x^2\,dx + \int_3^{12}(12 - x)\,dx$

$$= \left[\dfrac{x^3}{3}\right]_0^3 + \left[12x - \dfrac{x^2}{2}\right]_3^{12}$$

$$= 9 + 40.5 = \textbf{49.5}$$

Area $= \int_0^3 x^2\,dx + \dfrac{1}{2} \times 9 \times 9$

$$= 9 + 40.5 = \textbf{49.5}$$

You first need to find where the two curves intersect, by equating the two equations.

The graph shows the curve and the line.

You need to evaluate two integrals to find the area.

Alternatively, you can find the area under the curve and add the area of the triangle.

Q5 $x^2 - 5x + 9 = 3$

$x^2 - 5x + 6 = 0$

$x = 2$ or $x = 3$

Area $= 3 \times 1 - \int_2^3 (x^2 - 5x + 9)\,dx$

$$= 3 - \left[\dfrac{x^3}{3} - \dfrac{5x^2}{2} + 9x\right]_2^3$$

$$= 3 - \left(\dfrac{27}{2} - \dfrac{32}{3}\right)$$

$$= \dfrac{1}{6}$$

You must solve this equation to find the limits of integration.

The region is shown in the diagram.

The area is found by subtracting the value of the integral from the area of the rectangle with vertices at the points $(2, 3)$, $(3, 3)$, $(3, 0)$ and $(2, 0)$.

(b) $\dfrac{dA}{dx} = 48 - 8x$

$48 - 8x = 0 \Rightarrow x = 6$

x	5	6	7
$\dfrac{dy}{dx}$	8	0	-8

$A = (2 \times 6 + 5)(29 - 2 \times 6)$

$$= \textbf{289 m}^2$$

You should first expand the brackets and find the derivative.

To find the value of x you must solve the equation $\dfrac{dy}{dx} = 0$.

You can use this table to show that gradient changes from positive to negative at $x = 6$, so that you have a local maximum.

You can substitute $x = 6$ into the formula for A to find the maximum value.

Q9

(a) Perimeter $= 2x + 2y + 2y + \pi x$

$$= \textbf{2}x + \textbf{4}y + \pi x$$

You find the perimeter by considering a semicircle and three of the sides of the rectangle.

(b) Area $= 2x \times 2y + \dfrac{1}{2}\pi x^2$

$$= \textbf{4}xy + \dfrac{1}{2}\pi x^2$$

You find the area by considering a semicircle and a rectangle.

(c) $100 = 2x + 4y + \pi x$

$y = \dfrac{100 - 2x - \pi x}{4}$

$A = 4x\left(\dfrac{100 - 2x - \pi x}{4}\right) + \dfrac{1}{2}\pi x^2$

$$= \textbf{100}x - \textbf{2}x^2 - \dfrac{1}{2}\pi x^2$$

You should use the equation for the perimeter to express y in terms of x.

Then you can substitute for y in the expression for the area and simplify it to obtain the required result.

You could also collect the x^2-terms and write it as $100x - (2 + \frac{1}{2}\pi)x^2$.

(d) $\dfrac{dA}{dx} = 100 - 4x - \pi x$

$0 = 100 - 4x - \pi x$

$x = \dfrac{100}{4 + \pi} \Rightarrow \boldsymbol{y = 7.0\,m}$

You should form and solve the equation $\dfrac{dA}{dx} = 0$.
It is often a good idea to leave answers in the exact form like this.

You can substitute the value of x into the formula for y to find the corresponding value for y.

(e) $A = \dfrac{5000}{4 + \pi}$

$$= \textbf{700 m}^2 \ (2\ \text{s.f.})$$

Then substitute a value for x in the expression for A to obtain the actual area.

Q10 $500 = x^2h \Rightarrow h = \dfrac{500}{x^2}$

$$\boldsymbol{S = x^2 + 4xh = x^2 + \dfrac{2000}{x}}$$

$\dfrac{dS}{dx} = 2x - 2000x^{-2}$

$\dfrac{dS}{dx} = 0 \Rightarrow x = 10$

If $x = 9$, $\dfrac{dS}{dx} = -6.7$ and if

$x = 11$, $\dfrac{dS}{dx} = 5.5$.

First, you should consider the volume of the box, to express h in terms of x. Then you can obtain an expression for the surface area and substitute for h using the result above.

To find the minimum surface area, you should solve the equation $\dfrac{dS}{dx} = 0$.

You also need to show that the gradient changes from being negative to positive to check that you have a local minimum.

Q6 Area $= \int_{-2}^{1}(16 - x^4)\mathrm{d}x - \frac{1}{2} \times 3 \times 15$

$= \left[16x - \frac{x^5}{5}\right]_{-2}^{1} - 22.5$

$= 15.8 - (-25.6) - 22.5$

$= \mathbf{18.9}$

The diagram shows the area that must be found.

The area is given by the integral *minus* the area of the triangle between the line and the x-axis.

Q7 Area under the curve

$= \int_{1}^{4}(5x^2 - x^3)\mathrm{d}x - \frac{1}{2} \times 3 \times (4 + 16)$

$= \left[\frac{5x^3}{3} - \frac{x^4}{4}\right]_{1}^{4} - 30$

$= \frac{128}{3} - \frac{17}{12} - 30$

$= \frac{45}{4}$

The diagram shows the area that must be found.

The area is given by the integral *minus* the area of the trapezium between the line and the x-axis.

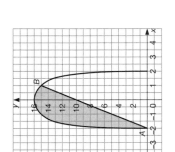

Q8 $\int_{0}^{2}(x^3 - 5x^2 + 6x)\mathrm{d}x$

$= \left[\frac{x^4}{4} - \frac{5x^3}{3} + 3x^2\right]_{0}^{2} = \frac{8}{3}$

$\int_{2}^{3}(x^3 - 5x^2 + 6x)\mathrm{d}x$

$= \left[\frac{x^4}{4} - \frac{5x^3}{3} + 3x^2\right]_{2}^{3}$

$= \frac{9}{4} - \frac{8}{3} = \frac{5}{12}$

Area $= \frac{8}{3} + \frac{5}{12} = \frac{37}{12}$

You should find the area of each of the two regions.

As the first region is above the axis you obtain a positive result from the integration.

As the second region is below the axis, when you integrate. The area shaded below the x-axis is $\frac{5}{12}$.

Finally you should calculate the total area.

Q9 (a) $\int_{0}^{4} x(4 - x)\mathrm{d}x$

$= \int_{0}^{4}(4x - x^2)\mathrm{d}x = \left[2x^2 - \frac{x^3}{3}\right]_{0}^{4}$

$= 2 \times 4^2 - \frac{4^3}{3} = \frac{32}{3}$

(b) (i) Area under curve

$= \int_{0}^{k} x(4 - x)\mathrm{d}x$

$= \int_{0}^{k}(4x - x^2)\mathrm{d}x$

$= \left[2x^2 - \frac{x^3}{3}\right]_{0}^{k}$

$= 2k^2 - \frac{k^3}{3}$

Area of triangle

$= \frac{1}{2} \times k \times k(4 - k)$

$= 2k^2 - \frac{k^3}{2}$

Area shaded

$= \left(2k^2 - \frac{k^3}{3}\right) - \left(2k^2 - \frac{k^3}{2}\right)$

$= \frac{k^3}{6}$

(ii) $\frac{k^3}{6} = \frac{1}{2} \times \frac{32}{3}$

$k = \sqrt[3]{32}$

$= \mathbf{3.175}$

You need to expand the brackets before integrating.

Use this integral to calculate the total area under the curve between $x = 0$ and $x = k$.

You should then express the area of the triangle in terms of k.

Finally you can find the difference to give the shaded area.

From part (a), you know that the total area under the curve from $x = 0$ to $x = 4$ is $\frac{32}{3}$. You also know that the area shaded is $\frac{k^3}{6}$. Form and solve this equation to find k.

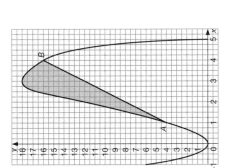

Q10 (a) $\frac{\mathrm{d}y}{\mathrm{d}x} = 2x - 2$

$2x - 2 = 0 \Rightarrow x = 1$

$y = 1^2 - 2 \times 1 + 2 = 1$

Coordinates are (1, 1).

(b) $4 - x = x^2 - 2x + 2$

$x^2 - x - 2 = 0 \Rightarrow x = -1$ or $x = 2$

Coordinates are **(−1, 5)** and **(2, 2).**

You need to first differentiate and then solve the equation $\frac{\mathrm{d}y}{\mathrm{d}x} = 0$. Then you can substitute $x = 1$ in the expression for y to find the y-coordinate.

You need to solve this equation to find the x-coordinate.

Then you can substitute your x-values into the equation for y to find the y-coordinates.

How to solve these questions

(c) Gradient of MQ $= \dfrac{1}{1} = 1$

Gradient of PQ $= -\dfrac{3}{3} = -1$

The product of these is -1, so MQ and PQ are perpendicular.

Area $= \dfrac{1}{2} \times$ PQ $\times$ MQ

$= \dfrac{1}{2} \times \sqrt{18} \times \sqrt{2} = \mathbf{3}$

You must first find the gradient of the two perpendicular sides. You may need to do this for all three sides.

You must show that the product of the gradients of two sides is -1.

Once you have shown that the sides PQ and MQ are perpendicular, you can use them as the base and the height of the triangle.

(d) Area shaded

$= \dfrac{1}{2} \times 3 \times (5+2) - \displaystyle\int_{-1}^{2}(x^2 - 2x + 2)\,\mathrm{d}x$

$= 10.5 - \left[\dfrac{x^3}{3} - x^2 + 2x\right]_{-1}^{2} = \mathbf{4.5}$

Area $= \dfrac{1}{2} \times$ area of triangle

Find the shaded area by taking away the area under the curve, given by the integral, from the area of the trapezium between the line and the x-axis.

Then compare this area to the area of the triangle.

8 ALGEBRA AND EQUATIONS II

Q1 $f(x) = x^4 - x^2 - 2x + 2$

$f(1) = 1 - 1 - 2 + 2 = 0$

$(x - 1)$ is a factor of $f(x)$.

$f(x) = (x - 1)(x^3 + x^2 - 2)$

Let $g(x) = x^3 + x^2 - 2$

$g(1) = 1 + 1 - 2 = 0$

$(x - 1)$ is a factor of $g(x)$.

$g(x) = (x - 1)(x^2 + 2x + 2)$

$f(x) = \mathbf{(x - 1)^2(x^2 + 2x + 2)}$

You need to do this in two steps.

First see if $(x - 1)$ is a factor of $f(x)$; if it is then factorise $f(x)$.

Now use the factor theorem to see if $(x - 1)$ is a factor of $x^3 + x^2 - 2$.

This is the final step since $x^2 + 2x + 2 = (x + 1)^2 + 1 > 0$ and cannot be factorised.

Q2 $f(1) = 1^3 - 7 \times 1 + 6$

$= 0 \Rightarrow x - 1$ is a factor.

$f(x) = (x - 1)(x^2 + x - 6)$

$= \mathbf{(x - 1)(x - 2)(x + 3)}$

First you evaluate $f(x)$ when $x = 1$.

If $f(1) = 0$ then $(x - 1)$ is a factor.

Now you take out the factor $(x - 1)$ and then factorise the quadratic.

Q3

(a) $f(-3) = 2(-3)^3 + 5(-3)^2 - 8(-3) - 15 = 0$

$\Rightarrow \mathbf{(x + 3)}$ **is a factor of** $\mathbf{f(x)}$.

(b) $\mathbf{f(x) = (x + 3)(2x^2 - x - 5)}$

(c) $2x^2 - x - 5 = 0 \Rightarrow x = \dfrac{1 \pm \sqrt{1 + 40}}{4}$

$\Rightarrow \mathbf{x = 1.85}$ **or** $\mathbf{x = -1.35}$

You can use the formula to solve the quadratic equation.

Answers

How to solve these questions

Q4

$$
\begin{array}{r}
x^2 + 2x - 3 \\
x^2 + 2 \overline{)\,x^4 + 2x^3 - x^2 + x - 4} \\
\underline{x^4 \qquad\quad + 2x^2} \\
2x^3 - 3x^2 + x - 4 \\
\underline{2x^3 \qquad\quad + 4x} \\
-3x^2 - 3x - 4 \\
\underline{-3x^2 \qquad\quad - 6} \\
-3x + 2
\end{array}
$$

The quotient is $x^2 + 2x - 3$ and the remainder is $-3x + 2$.

You must be careful to subtract the corresponding powers at each of these steps.

Q5

$\dfrac{(3x + 1)(x - 1)(x + 2)}{(x^2 + x + 1)(x - 1)(x + 2)} + \dfrac{2(x + 2)(x^2 + x + 1)}{(x^2 + x + 1)(x - 1)(x + 2)}$

$- \dfrac{3(x - 1)(x^2 + x + 1)}{(x^2 + x + 1)(x - 1)(x + 2)}$

$= \dfrac{(3x^3 + 4x^2 - 5x - 2) + (2x^3 + 6x^2 + 6x + 4) - (3x^3 - 3)}{(x^2 + x + 1)(x - 1)(x + 2)}$

$= \dfrac{\mathbf{2x^3 + 10x^2 + x + 5}}{\mathbf{(x^2 + x + 1)(x - 1)(x + 2)}}$

First you need to find the common denominator since $(x^2 + x + 1)$ cannot be factorised.

$(x - 1)$ and $(x + 2)$ are therefore not factors. The common denominator is $(x^2 + x + 1)(x - 1)(x + 2)$.

Now write each expression over the common denominator and combine the numerator.

Q6

(a) If $x_n \to L$ as $n \to \infty$ then $x_{n+1} = x_n = L$

$L = \dfrac{2L^3 + a}{3L^2}$

$L^3 = a$

$\mathbf{L = \sqrt[3]{a}}$

(b) For $\sqrt[3]{3}$, $a = 3$

$x_{n+1} = \dfrac{2x_n^3 + 3}{3x_n^2}$

$x_0 = 1$

$x_1 = 1.6667$

$x_2 = 1.4711$

$x_3 = 1.4428$

$x_4 = 1.4422$

$x_5 = 1.4422$ to three decimal places

$\sqrt[3]{3} = \mathbf{1.442}$

Remember that after many steps $x_{n+1} = x_n$, if the sequence converges.

Substitute $x_{n+1} = x_n = L$ and solve for L.

You can stop the iteration when two answers are equal to two decimal places.

Answers

Q7

$|x + 3| < 2|x| \Leftarrow x > 3$

You cannot use ⇒ or ⇔ because x < −1 is also a solution.

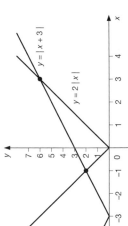

How to solve these questions

Draw a sketch of the two functions $y = |x + 3|$ and $y = 2|x|$.
You see that $|x + 3| = 2|x|$ when $x = 3$ and when $x = −1$.
The graph shows that $|x + 3| < 2|x|$ for $x < −1$ and for $x > 3$. Explain why you cannot use ⇒ or ⇔.

9 EXPONENTIALS AND LOGARITHMS

Q1

(a) $f(0) = 4^0 = $ **1**

(b) $f(x + y) = 4^{x+y} = 4^x 4^y = $ **f(x)f(y)**

(c) $\dfrac{f(x)}{f(y)} = \dfrac{4^x}{4^y} = 4^{x-y} = $ **f(x−y)**

(d) $\{f(x)\}^n = (4^x)^n = 4^{nx} = $ **f(nx)**

Q2

(a) $\dfrac{1}{3} \ln e^3 = \dfrac{1}{3}(3 \ln e) = \ln e = $ **1**

(b) $6 \ln \sqrt{e} = 6 \ln e^{\frac{1}{2}} = 6 \times \dfrac{1}{3} \ln e = $ **2**

(c) $7 \ln e^4 - 6 \ln \sqrt{e} = 28 \ln e - 3 \ln e$
$= $ **25**

(d) $e^{4 \ln x} = e^{\ln x^4} = $ **x^4**

(e) $\ln(e^{2x}) = $ **2x**

Q3

(a) $P_0 = 50\,000 e^{0.05 t_0}$
$2P_0 = 50\,000 e^{0.05 t}$
$2 = e^{0.05(t_1 - t_0)}$
$t_1 - t_0 = \dfrac{1}{0.05} \ln 2 = $ **13.86 years**

(b) $55\,000 = 50\,000 e^{0.05 t}$
$1.1 = e^{0.05 t} \Rightarrow 0.05 t = \ln 1.1$
$t = $ **1.91 years**

How to solve these questions

You should substitute in $f(x)$ for $x = 0$ and remember that $a^0 = 1$ for all $a \neq 0$.

You should remember that $\ln e = 1$ and $e^{\ln a} = a$.

You choose a population P_0 at time t_0.
You need to find t, for $P = 2P_0$.
Divide the two equations to eliminate P_0.
This is the time taken for the population to double.

Answers

Q4

Let $y = e^{2x}$
$y^2 + y - 6 = 0$
$(y + 3)(y - 2) = 0$
$y = -3$ (no solution) or $y = 2$
$e^{2x} = 2 \Rightarrow x = \dfrac{1}{2} \ln 2 = $ **0.3466**

Q5

(a) $\ln x = 7 \Rightarrow x = e^7 = $ **1096.6**

(b) $3 - 4 \ln x = 0$
$\Rightarrow \ln x = \dfrac{3}{4}$
$\Rightarrow x = e^{\frac{3}{4}} = $ **2.117**

(c) $3 \ln x^3 + 4 \ln x = 9$
$\Rightarrow 9 \ln x + 4 \ln x = 9$
$\Rightarrow \ln x = \dfrac{9}{13}$
$\Rightarrow x = e^{\frac{9}{13}} = $ **1.998**

(d) $\ln 7x + \ln 5x = 1$
$\ln 35x^2 = 1$
$35x^2 = e^1$
$x = $ **0.2787**

(e) $\ln (2x - 1) - \ln x = 0$
$\ln \dfrac{2x - 1}{x} = 0$
$\dfrac{2x - 1}{x} = 1$
x = 1

Q6

Let $N = 200\,000$ when $t = 0$.
Let $N = 300\,000$ when $t = 5$.
$\Rightarrow N_0 = 200\,000$
$300\,000 = 200\,000 e^{5\kappa}$
$\kappa = \dfrac{1}{5} \ln 1.5 = 0.081\,09$
$N = 200\,000 e^{10\kappa}$
$= $ **450 000**

How to solve these questions

First you need to form an equation without exponentials.
Now you solve the quadratic in y.
For $y = -3$ there are no solutions for x.

Now you simplify $\ln x^3 = 3 \ln x$.
The only solution is for $x > 0$.

First you need to find the values of N_0 and κ.

Now you put $t = 10$ to find N after 10 hours.

Q7

(a) $f'(x) = e^x - 5$

(b) $f' = 0 \Rightarrow e^x - 5 = 0$

$e^x = 5 \Rightarrow x = \ln 5 = \mathbf{1.609}$

(c) $f(0.2) = 0.22$

$f(0.3) = -0.15$

$\therefore f(x) = 0$ for $\mathbf{0.2 < \alpha < 0.3}$

(d) From a graph of $y = e^x - 5x$ or a table of values there is a change of sign for $P = 25$ i.e. **between $x = 2.5$ and $x = 2.6$.**

First, find $f'(x)$.

For a decreasing function, form and solve the inequality $f'(x) < 0$.

Differentiate twice to obtain the first and second derivatives.

You find the gradient by finding the derivative, so substitute $x = 2$ into the expression for $\dfrac{dy}{dx}$.

If α is a root between 0.2 and 0.3 then the function changes sign.

10 DIFFERENTIATION II

Q1

(a) $\dfrac{dy}{dx} = 4x^3 + 6x^5$

When $x = 0$, $\mathbf{\dfrac{dy}{dx} = 0}$

$\dfrac{d^2y}{dx^2} = 12x^2 + 30x^4$

When $x = 0$, $\mathbf{\dfrac{d^2y}{dx^2} = 0}$

(b) At $x = -1$, $\dfrac{dy}{dx} = -10$

and $x = 1$, $\dfrac{dy}{dx} = 10$, **so there is a local minimum at $x = 0$.**

Q2

$\dfrac{dy}{dx} = a - \dfrac{1}{x}$

$a - \dfrac{1}{x} = 0 \quad$ so $x = \dfrac{1}{a}$

$\dfrac{d^2y}{dx^2} = \dfrac{1}{x^2}$

When $x = \dfrac{1}{a}$, $\dfrac{d^2y}{dx^2} = a^2$.

∴ a local minimum

Q3

$f'(x) = 1 - 2e^{-x}$

$1 - 2e^{-x} < 0$

$e^{-x} > \dfrac{1}{2}$

$\boldsymbol{x < \ln 2}$

Q4

(a) $\dfrac{dy}{dx} = -\dfrac{1}{x^2} + \dfrac{1}{x}$

$\dfrac{d^2y}{dx^2} = \dfrac{2}{x^3} - \dfrac{1}{x^2}$

(b) When $x = 2$, $\dfrac{dy}{dx} = \dfrac{1}{4}$

$\therefore$ **gradient is $\dfrac{1}{4}$.**

(c) $\dfrac{dy}{dx} = -\dfrac{1}{x^2} + \dfrac{1}{x}$

$\dfrac{dy}{dx} = 0 \Rightarrow 0 = x - 1$

$x = 1$

At $x = 1$, $\dfrac{d^2y}{dx^2} = 1$

Local minimum at $(1, 1 + \ln 2)$.

Q5

(a) $\dfrac{dy}{dx} = 2 - 6e^{-2x}$

$2 - 6e^{-2x} = 0$

$e^{-2x} = \dfrac{1}{3}$

$x = \dfrac{1}{2}\ln 3$

Stationary point at $\left(\dfrac{1}{2}\ln 3, \ln 3 + 1\right)$.

(b) $\dfrac{d^2y}{dx^2} = 12e^{-2x}$

(c) When $x = \dfrac{1}{2}\ln 3$, $\dfrac{d^2y}{dx^2} = 4$

∴ a local minimum.

First, find $f'(x)$.

For a decreasing function, form and solve the inequality $f'(x) < 0$.

Differentiate twice to obtain the first and second derivatives.

You find the gradient by finding the derivative, so substitute $x = 2$ into the expression for $\dfrac{dy}{dx}$.

For a stationary point $\dfrac{dy}{dx} = 0$, so form and solve this equation. There is a stationary point at $x = 1$. To determine the nature of the stationary point look at the second derivative.

As $\dfrac{d^2y}{dx^2} > 0$ you can conclude that there is a local minimum when $x = 1$. Finally, substitute $x = 1$ to obtain the y-coordinate.

For a stationary point $\dfrac{dy}{dx} = 0$, so first you need to differentiate.

Then form and solve the equation $\dfrac{dy}{dx} = 0$, to find the x-coordinate of the stationary point.

Substitute the x-value into the expression for y to find the y-coordinate.

Differentiate $\dfrac{dy}{dx}$ again to find the second derivative.

As $\dfrac{d^2y}{dx^2} > 0$ the stationary point is a local minimum.

First differentiate, and then substitute $x = 0$. Note that there must be a stationary point at $x = 0$.

Differentiate again to find the second derivative.

As this is zero you cannot determine the nature of the stationary point from this information.

To determine the nature of the stationary point you should consider the gradient of the curve either side of $x = 0$, in this case we have used $x = -1$ and $x = 1$.

As the gradient changes from being negative to positive, the stationary point must be a local minimum.

First you should differentiate.

To find the stationary point, form and solve the equation $\dfrac{dy}{dx} = 0$.

Use the second derivative to determine the nature of the stationary point by substituting $x = \dfrac{1}{a}$.

As $\dfrac{d^2y}{dx^2} > 0$, the stationary point must be a local minimum.

Q6

(a) $\frac{dy}{dx} = \frac{1}{x} - 2x$

and

$\frac{d^2y}{dx^2} = -\frac{1}{x^2} - 2$

Differentiate once, and then differentiate again noting that $\frac{1}{x} = x^{-1}$.

(b) $0 = \frac{1}{x} - 2x$

$1 - 2x^2 = 0$

$x = \frac{1}{\sqrt{2}}$

When $x = \frac{1}{\sqrt{2}}$, $\frac{d^2y}{dx^2} = -4$

∴ **a local maximum at**

$\left(\frac{1}{\sqrt{2}}, \ln(2\sqrt{2}) - \frac{1}{2}\right)$

You need $\frac{dy}{dx} = 0$ for a stationary point, so form and solve this equation. Note that you can't use the negative root as $\ln(4x)$ is not defined for negative x.

As $\frac{d^2y}{dx^2} < 0$, you have a local maximum.

Substituting $x = \frac{1}{\sqrt{2}}$ into the expression for y gives you the y-coordinate.

Q7

$\frac{dy}{dx} = 1 + \frac{1}{x}$

At $x = 1$, this gives $\frac{dy}{dx} = 2$.

∴ the equation of the tangent is $y = 2x + c$

When $x = 1$, $y = 1 + \ln 2$

$1 + \ln 2 = 2 + c$

$c = \ln 2 - 1$

So the equation of the tangent is $y = 2x + \ln 2 - 1$.

First you should differentiate.

Substituting $x = 1$ gives the gradient of the curve at this point. This will be the gradient of the tangent.

You can now substitute the gradient into the equation of a straight line $y = mx + c$.

Now you need to find the value of the constant c.

Substitute these values into the equation $y = x + c$ and then solve for c.

Once you have found c, you can write down the equation of the tangent.

Q8

(a) $\frac{dy}{dx} = -2e^{2x}$

At the point A, $x = 0$.

When $x = 0$, $\frac{dy}{dx} = -2$

The equation of the tangent is $y = -2x + c$.

The coordinates of A are (0, 1).

$1 = 0 + c$

$c = 1$

The equation of the tangent is $y = -2x + 1$.

First, differentiate.

Note that $x = 0$, as A lies on the y-axis.

Next, substitute $x = 0$, to find the gradient of the tangent.

Now substitute the gradient into the equation of a straight line $y = mx + c$.

Substitute $x = 0$ in the expression for y to find the y-coordinate of the point A.

Substituting $x = 0$ and $y = 1$, allows you to find the value of c.

Now substitute $c = 1$, to complete the equation of the tangent.

(b) The normal will have gradient $m = \frac{-1}{-2} = \frac{1}{2}$

$y = \frac{1}{2}x + c$

The coordinates of A are (0, 1).

$1 = 0 + c$

$c = 1$

The equation of the normal is $y = \frac{1}{2}x + 1$.

$\frac{1}{2}x + 1 = 0$

$x = -2$

The coordinates are **(–2, 0)**.

As the normal is perpendicular to the tangent at the same point, you can use:

gradient of normal $= -\dfrac{1}{\text{gradient of tangent}}$

Substitute $m = \frac{1}{2}$ into the equation of a straight line.

You should substitute these values to find c.

Next you should substitute this value of c to complete the equation.

The normal intersects the x-axis, when $y = 0$, so you should form and solve the equation $y = 0$.

You can substitute $y = -2$ into the expression for y to find the y-coordinate.

Q9

$\frac{dy}{dx} = -\frac{2}{x}$

At $x = e$ the gradient of the curve will be $-\frac{2}{e}$ and so the gradient of the normal will be $\frac{e}{2}$ and the equation of the normal is

$y = \frac{e}{2}x + c$.

When $x = e$, $y = 3$.

$3 = \frac{e^2}{2} + c$

$c = 3 - \frac{e^2}{2}$

The equation of the normal is $y = \dfrac{e}{2}x + 3 - \dfrac{e^2}{2}$.

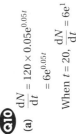

First, differentiate y to obtain $\frac{dy}{dx}$.

Substituting $x = e$ will give you the gradient of curve and the tangent. Then you can use:

gradient of normal $= -\dfrac{1}{\text{gradient of tangent}}$

Then substitute the gradient into the equation of a straight line.

You can find the y-value by substituting $x = e$ into the expression for y.

You can then find c by substituting these values into the equation.

Q10

(a) $\frac{dN}{dt} = 120 \times 0.05e^{0.05t}$

$= 6e^{0.05t}$

When $t = 20$, $\frac{dN}{dt} = 6e^1$

$= \mathbf{16.3}$ (3 s.f.)

The rate of increase is given by the derivative, $\frac{dN}{dt}$.

Substitute $t = 20$ in $\frac{dN}{dt}$, to find the required rate.

Answers

(b) $360 = 120e^{0.05t}$

$3 = e^{0.05t}$

$t = 20\ln3$

$\dfrac{dN}{dt} = 6e^{0.05 \times 20\ln3}$

$= 6e^{\ln3}$

$= \mathbf{18}$

11 INTEGRATION II

Q1 Area $= \displaystyle\int_0^3 (e^x - 2x)\,dx$

$= \left[e^x - x^2\right]_0^3$

$= (e^3 - 9) - e^0$

$= \mathbf{e^3 - 10}$

Q2 Area $= \displaystyle\int_0^2 (e^{3x} + 1)\,dx$

$= \left[\dfrac{1}{3}e^{3x} + x\right]_0^2$

$= \left(\dfrac{1}{3}e^6 + 2\right) - \left(\dfrac{1}{3}e^0 + 0\right)$

$= \mathbf{\dfrac{1}{3}(e^6 + 5)}$

Q3 Area $= \dfrac{1}{2} \times 1 \times 1 + \displaystyle\int_0^3 e^{-x}\,dx$

$= \dfrac{1}{2} + \left[-e^{-x}\right]_0^3$

$= \dfrac{1}{2} - e^{-3} + 1$

$= \mathbf{\dfrac{3}{2} - e^{-3}}$

How to solve these questions

First form and solve the equation $N = 360$, so that you find the corresponding value of t.

Then substitute this value of t to find the rate.

Use k as the upper limit of integration.

Note that the limits of integration are $x = 0$ and $x = 3$.

Note that the limits of integration are $x = 0$ and $x = 2$.

The diagram shows the region.

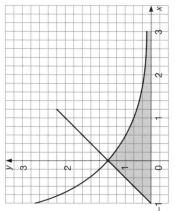

Answers

Q4 $y = \displaystyle\int 2e^{\frac{1}{2}x}\,dx = 4e^{\frac{1}{2}x} + c$

$0 = 4e^{\ln3 - \ln2} + c$

$c = -6$

The equation of the curve is

$\mathbf{y = 4e^{\frac{1}{2}x} - 6.}$

Q5 Volume $= \displaystyle\int_1^4 \pi\left(\dfrac{1}{x}\right)^2 dx$

$= \displaystyle\int_1^4 \pi x^{-2}\,dx$

$= \left[-\pi\left(\dfrac{1}{x}\right)\right]_1^4$

$= \mathbf{\dfrac{3\pi}{4}}$

Q6 Volume $= \displaystyle\int_1^4 \pi(\sqrt{x^2 + 1})^2\,dx$

$= \pi\displaystyle\int_1^4 (x^2 + 1)\,dx$

$= \pi\left[\dfrac{x^3}{3} + x\right]_1^4$

$= \pi\left[\dfrac{76}{3} - \dfrac{4}{3}\right]$

$= \mathbf{24\pi}$

Q7 $\sqrt{x - 1} = 0$

$x = 1$

Volume $= \displaystyle\int_1^3 \pi(\sqrt{x - 1})^2\,dx$

$= \pi\displaystyle\int_1^3 (x - 1)\,dx$

$= \pi\left[\dfrac{x^2}{2} - x\right]_1^3$

$= \pi\left[\dfrac{3}{2} - \left(-\dfrac{1}{2}\right)\right]$

$= \mathbf{2\pi}$

How to solve these questions

Integrate the expression for the gradient to obtain the equation of the curve. You must include a constant of integration.

You should use the coordinates of the point on the curve to find c.

You should use the formula for the volume of a solid formed by rotation about the x-axis.

Use the formula for the volume of a solid formed by rotation about the x-axis.

You first need to find lower limit of integration solving the equation $y = 0$.

Then you can use the standard formula to find the volume.

Note that the area is made up of a triangle, for which the area is easy to calculate, and a region bounded by a curve, for which the area must be found using integration.

Q8 $x - x^2 = 0$

$x = 0$ and $x = 1$.

Volume $= \int_0^1 \pi(x - x^2)^2 dx$

$= \pi \int_0^1 (x^2 - 2x^3 + x^4) dx$

$= \pi \left[\frac{x^3}{3} - \frac{x^4}{2} + \frac{x^5}{5} \right]_0^1$

$= \pi \left[\frac{1}{3} - \frac{1}{2} + \frac{1}{5} \right]$

$= \frac{\pi}{30}$

First, solve the equation $y = 0$, to obtain the limits of integration.

Then you can find the volume, using a standard integral.

Q9 Area $= \int_{-2}^0 \frac{1}{2} e^x dx$

$= \left[\frac{1}{2} e^x \right]_{-2}^0$

$= \frac{1}{2} - \frac{1}{2} e^{-2}$

$= \frac{1}{2} \left(1 - \frac{1}{e^2} \right)$

Note that the lower limit of integration is –2 and that the upper limit is 0.

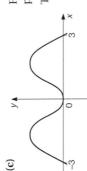

Q10 $y = \int \left(2e^{2x} + 5 \right) dx$

$= e^{2x} + 5x + c$

$5 + 5\ln 2 = e^{2\ln 2} + 5\ln 2 + c$

$5 = e^{\ln 4} + c$

$5 = 4 + c$

$c = 1$

$f(x) = e^{2x} + 5x + 1$

Obtain the equation of the curve by integrating $\frac{dy}{dx}$.

Substitute the values of x and y at the point for which the coordinates are given.

Finally, put in the value of c to find f(x).

Q11 $\int_0^k e^{-3x} dx = \frac{7}{24}$

$\left[-\frac{1}{3} e^{-3x} \right]_0^k = \frac{7}{24}$

$-\frac{1}{3} e^{-3k} + \frac{1}{3} = \frac{7}{24}$

$e^{-3k} = \frac{1}{8}$

$k = -\frac{1}{3} \ln \left(\frac{1}{8} \right)$

$= \ln 2$

As you know that the integral is $\frac{7}{24}$ you can form and solve an equation for k.

Note that the answer is given here in an exact form and not as a decimal.

12 TRANSFORMATIONS

Q1 **Translate 2 units to the right. Stretch, scale factor $\frac{1}{4}$ parallel to the x-axis.**

Q2

(a)

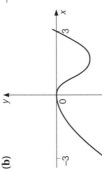

f($-x$) is a reflection in the y-axis.

(b)

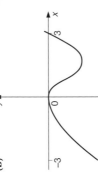

$-$f(x) is a reflection in the x-axis.

(c)

For f($|x|$) you need to remember that for all values of x, $|x|$ is positive. So for example, if $x = -2$ then $y = $ f($|-2|$) $= $ f(2).

The portion of f for $x > 0$ is reflected in the y-axis.

Answers

Q3

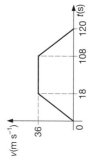

(a) (b) $y = 2$ (c)

$x = -1$

$$\frac{2x+3}{x+1} = \frac{2(x+1)+1}{x+1} = 2 + \frac{1}{x+1}$$

The graph is symmetrical about $(-1, 2)$.

$\frac{2x+3}{x+1}$ is a **translation**, $\begin{pmatrix} -1 \\ 2 \end{pmatrix}$

Q4

(a)

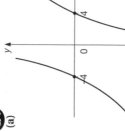

(b)

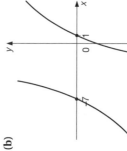

How to solve these questions

For (b), you translate the graph 1 unit to the left, parallel to the x-axis.

For (c), you translate the graph in (b) up 2 units parallel to the y-axis.

From part (c) you can deduce that $\frac{2x+3}{x+1}$ is a translation of 1 unit to the left and 2 units up.

For an odd function $f(-x) = -f(x)$ the graph for $x < 0$ is a rotation through a half-turn about $(0, 0)$.

$f(x + 3)$ translates the graph to the left by 3 units.

The points of intersection with the x-axis are $(1, 0)$ and $(-7, 0)$.

Answers

13 KINEMATICS ON A STRAIGHT LINE

Q1

(a) $0^2 = 6^2 + 2 \times (-9.8)s$

$s = \frac{36}{2 \times 9.8} = 1.84\,\text{m}$

Maximum height $= 5 + 1.84$
$= \mathbf{6.84\,m}$

(b) $-5 = 6t + \frac{1}{2}(-9.8)t^2$

$4.9t^2 - 6t - 5 = 0$

$t = 1.79$ or $t = -0.57$

Time in air = **1.79 s**

(c) $v^2 = 6^2 + 2 \times (-9.8) \times (-5)$

$v = \sqrt{134} = \mathbf{11.58\,m\,s^{-1}}$

Q2

(a) $v = 10 + 3 \times 6$
$= \mathbf{28\,m\,s^{-1}}$

(b) For OA:

$10^2 = 0^2 + 2 \times 4 \times s$

$s = 12.5\,\text{m}$

For AB:

$s = 10 \times 6 + \frac{1}{2} \times 3 \times 6^2$

$= 114\,\text{m}$

Total distance = 12.5 + 114
$= \mathbf{126.5\,m}$

Q3

(a)

$v\,(\text{m s}^{-1})$

36

0   18   108   120 $t(s)$

(b) Distance $= \frac{1}{2}(90 + 120) \times 36$
$= \mathbf{3780\,m}$

How to solve these questions

First note that $a = -9.8$ and $v = 0$, at the maximum height of the ball.

Then use $v^2 = u^2 + 2as$.

You must add on the height from which the ball was thrown.

Note that when the ball hits the ground, $s = -5$ and use the equation $s = ut + \frac{1}{2}at^2$.

You must take the positive solution in this case.

Use $s = -5$ and the equation $v^2 = u^2 + 2as$.

Use $v = u + at$ with $u = 10$, $a = 3$ and $t = 6$ to find the speed.

Calculate the distance for each part.

You should use $v^2 = u^2 + 2as$.

Here you should use $s = ut + \frac{1}{2}at^2$.

Add the two distances to find OB.

Your graph should show each stage of the journey and you should mark in the key times and the velocity as shown here.

You can find the distance by using the graph and finding the area.

(c) It travels at constant speed for most of its journey.

This is the most obvious difference.

(d) $3780 = \frac{1}{2} \times 150 \times V$

$V = 50.4 \text{ m s}^{-1}$

The graph is in the shape of a triangle, which has height V and base 150.

Q4

(a) **The train slows down, then travels at a constant speed before slowing down and coming to rest.**

You should identify the key features of the journey.

(b) 1st stage:

distance $= \frac{1}{2}(3 + 1.2) \times 5$

$= 10.5 \text{ m}$

You should calculate the distance for each stage of the journey.

2nd stage:

distance $= 1.2 \times 5$

$= 6 \text{ m}$

3rd stage:

distance $= \frac{1}{2} \times 6 \times 1.2$

$= 3.6 \text{ m}$

For the first stage you need to find the area of a trapezium, for the second stage the area of a rectangle and for the third stage the area of a triangle.

Total distance $= 10.5 + 6 + 3.6$

$= 20.1 \text{ m}$

Then you can find the total distance.

(c) At $t = 12$, $v = 0.8 \text{ m s}^{-1}$

From $t = 10$ to 12:

$s = \frac{1}{2}(1.2 + 0.8) \times 2 = 2 \text{ m}$

First, calculate the speed when $t = 12$.

Then you can use $s = \frac{1}{2}(u + v)t$.

Total distance $= 10.5 + 6 + 2$

$= 18.5 \text{ m}$

Again you will need to include the distance travelled in the first two stages.

Q5

(a)

Your graph should have three stages.

Make sure you label all the key values on each axis.

(b) $15 = \frac{1}{2} \times 1.5 \times 4 + 4t$

$\qquad + \frac{1}{2} \times 1 \times 4$

$15 = 5 + 4t$

$t = 2.5 \text{ s}$

You can form this equation by noting that the total area under the graph must be 15.

Total time $= 1.5 + 2.5 + 1 = 5 \text{ s}$

Don't forget that you need the total time.

(c) Average velocity $= \frac{15}{5}$

$= 3 \text{ m s}^{-1}$

Use: average velocity $= \dfrac{\text{total displacement}}{\text{total time}}$

Q6

(a) Let V be the speed and T the time at B.

$V = 20 + 2T$

You should use $v = u + at$ here.

$AB = 20T + \frac{1}{2} \times 2T^2$

$= 20T + T^2$

You should use $s = ut + \frac{1}{2}at^2$ here.

$BC = 10V$

$= 10(20 + 2T)$

$= 200 + 20T$

Find the distance BC in terms of V and then convert this so that it is in terms of T.

$425 = 20T + T^2 + 200 + 20T$

$0 = T^2 + 40T - 225$

$T = 5$ or $T = -45$

Time to travel from A to B is 5 s.

You can then use the fact that the total distance is 425 m to form and solve an equation.

(b) $V = 20 + 2 \times 5$

$= 30 \text{ m s}^{-1}$

You can use your earlier expression for V with $T = 5$.

14 KINEMATICS AND VECTORS

Q1

$\mathbf{i} - 3\mathbf{j} = 4\mathbf{i} + 2\mathbf{j} + \mathbf{a} \times 10$

$10\mathbf{a} = -3\mathbf{i} - 5\mathbf{j}$

$\mathbf{a} = (-0.3\mathbf{i} - 0.5\mathbf{j}) \text{ m s}^{-2}$

You need to use the formula $\mathbf{v} = \mathbf{u} + \mathbf{a}t$ with $\mathbf{v} = \mathbf{i} - 3\mathbf{j}$, $\mathbf{u} = 4\mathbf{i} + 2\mathbf{j}$ and $t = 10$. Then solve to find $\mathbf{a}$.

Q2

(a) $\mathbf{r} = (5\mathbf{i} - 5\mathbf{j})t + \frac{1}{2}(-\mathbf{i} + 2\mathbf{j})t^2$

$= (5t - \frac{1}{2}t^2)\mathbf{i} + \frac{1}{2}(-5t + t^2)\mathbf{j}$

You should use the formula $\mathbf{r} = \mathbf{u}t + \frac{1}{2}\mathbf{a}t^2$ to find the position.

$\mathbf{v} = (5\mathbf{i} - 5\mathbf{j}) + (-\mathbf{i} + 2\mathbf{j})t$

$= (5 - t)\mathbf{i} + (2t - 5)\mathbf{j}$

You should use the formula $\mathbf{v} = \mathbf{u} + \mathbf{a}t$ to find the velocity.

Answers

(b) When $5 - t = 0$
$t = 5$ s

Q3

(a) $r_A = (4i + 3j)t + 4j$
$= 4ti + (3t + 4)j$
$r_B = (2i - j)t + 4i + 12j$
$= (2t + 4)i + (12 - t)j$

(b) $4t = 2t + 4$
$t = 2$
$3t + 4 = 12 - t$
$t = 2$
∴ they collide when $t = 2$
The position at collision is
$8i + 10j$.

(c) When $t = 1$:
$r_A = 4i + 7j$
$r_B = 6i + 11j$
$r_B - r_A = 2i + 4j$
Distance $= \sqrt{2^2 + 4^2}$
$= \mathbf{4.47\ m}$ (3 s.f.)

Q4

(a)

(b) $\dfrac{\sin\beta}{0.8} = \dfrac{\sin 30°}{1.2}$
$\beta = 19.47°$
$\alpha = 130.53°$
$v = \mathbf{1.82\ m\,s^{-1}}$ (3 s.f.)

How to solve these questions

When it is heading north the i component of the velocity will be zero. You should use this to form and solve an equation.

The positions can be found by multiplying the velocity by t and adding the initial position.

You need both components to be equal at the time of the collision. This gives two equations that will have the same solution if a collision takes place.

Then you can substitute this time into either position vector, to find where the boats collide.

You should first find the positions of both boats at this time. Find the displacement of B relative to A.

Use Pythagoras' theorem to calculate the distance.

Your triangle should show the current, which has a velocity of 0.8 m s⁻¹, the resultant velocity, which is at 30° to the current, and the velocity of the swimmer which is 1.2 m s⁻¹, and that completes the triangle.

You need to find the length of the longest side which represents the resultant velocity. First find the angle β, using the sine rule.

Then you can calculate α.

Finally you can use the sine rule again to find the resultant velocity.

Answers

Q5

(a) (i)
After 200 seconds:
$r = 200i \times 200 +$
$\frac{1}{2} \times(-0.5i - 0.05j) \times 200^2$
$- 50\,000i + 4000j$
$= -20\,000i + 3000j$

$v = 200i + (-0.5i - 0.05j) \times 200$
$= 100i - 10j$
Extra time $= \dfrac{3000}{10}$
$= 300$ seconds
The plane lands after 500 seconds.

(ii) $r = (-20\,000 + 100 \times 300)i = \mathbf{10\,000i}$

(iii) $v = \sqrt{100^2 + 10^2}$
$= \mathbf{100\ m\,s^{-1}}$ (3 s.f.)

(b) Not a good assumption, as the vertical component would probably be reducing to make the landing more comfortable

Q6

(a) $r = 20i$
$s = (10i + 10j)t + 300i$
$= \mathbf{(10t + 300)i + 10tj}$

(b) $\overrightarrow{AB} = s - r$
$= \mathbf{(300 - 10t)i + 10tj}$

(c) $300 - 10t = 10t$
$t = \mathbf{15\ s}$

(d) $(300 - 10t)^2 + (10t)^2 = 300^2$
$200t^2 - 6000t = 0$
$t = 0$ or 30
So they are 300 m apart again when $t = 30$.

How to solve these questions

First you should find the position and velocity after 200 seconds.

The plane has to descend 3000 m at a rate of 10 m s⁻¹.

This is the total time from the initial position.

You only need to consider the horizontal component, as the vertical one is zero.

You have to find the magnitude of the velocity found in part (i).

Make a statement and give a reason to support it.

The position vectors are given by the product of the velocity and t plus the initial positions.

This is found by subtracting the position of A from the position of B.

When on this bearing both components of AB will be equal.

Use Pythagoras' theorem to find the distance between A and B.

Then form and solve this quadratic equation.

15 NEWTON'S LAWS AND CONNECTED PARTICLES

Q1

(i) *See diagram opposite.*

(ii) It would not make any difference because the angle β remains the same wherever Brian goes. To support the packing case in the same position requires the two tensions T_1 in XY and ZY. **This is the tension in Brian's string.**

(iii) Resolving horizontally:
$F = T_2 \cos\alpha$ (1)
Resolving vertically:
$2T_1 \sin\beta = W + T_2 \sin\alpha$ (2)

(iv) $W = 200,\ F = 50,\ \alpha = 45°\ \beta = 75°$
From (1):
$T_2 = \dfrac{50}{\cos 45°} = 50\sqrt{2}\,N = 70.71\,N$
From (2):
$T_1 = \dfrac{200 + T_2 \sin 45°}{2\sin 75°}$
$= 129.41\,N$

(v) *See diagram opposite.*
Resolving vertically:
$R_1 = T_1 \sin 75° + T_1 \sin 30°$
$= 189.71\,N$
Resolving horizontally:
$R_2 = T_1 \cos 30° - T_1 \cos 75°$
$= 78.58\,N$
The magnitude of the reaction force is $\sqrt{R_1^2 + R_2^2}$
$= 205.37\,N$

How to solve these questions

The tension in Brian's rope is T_1 so the effect at Y is two tensions of T_1.

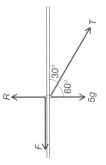

You start with a force diagram at the point X. It is often easier in these problems to write the unknown reaction force in component form.

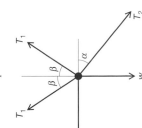

Answers

(vi) If the wind blows stronger then F increases in magnitude. If F increases in magnitude then for equilibrium to be maintained T_2 increases (equation 1) and hence T_1 increases (equation 2). **Brian must pull harder.**

Q2 *See diagram opposite.*

Resolving forces vertically:
$R = 5g + T\cos 60°$ (1)
Resolving forces horizontally:
$F = T\cos 30°$ (2)
Using the law of limiting friction:
$F = \dfrac{1}{2}R$ (3)
(3) and (2) ⇒ $T\cos 30° = \dfrac{1}{2}R$
$R = 2T\cos 30°$
From (1):
$5g + T\cos 60° = 2T\cos 30°$
$T = \dfrac{5g}{2\cos 30° - \cos 60°}$
$T = \mathbf{39.8\,N}$

How to solve these questions

Show the forces clearly.

Remember that the friction opposes the direction of the motion of the ring.

Since the ring is about to slip, use limiting friction $F = \mu R$.

Use $g = 9.8\ m\,s^{-2}$ here.

Answers

Q3

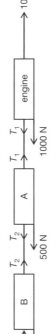

Using the law of friction:

$F = 0.2R$ (1)

Applying Newton's second law to the 2-kg mass:

$2g - T = 2a$ (2)

Applying Newton's second law to the 6-kg mass:

$R - 6g = 0$ (3)

$T - F = 6a$ (4)

From (3) and (1):

$F = 0.2R = 0.2 \times 6g$

$F = 1.2g$

Adding (2) and (4):

$2g - F = 8a$

$a = 0.1g$

From (2): $T = 2g - 2a = \mathbf{1.8g}$

Q4

First, draw a diagram and show all the forces on it.

How to solve these questions

First, draw a diagram and show all the forces on the objects.

Apply Newton's second law to each object in the system separately.

Answers

(i)

Locomotive:

$10\,000 - T_1 - 1000 = 50\,000a$ (1)

Truck A:

$T_1 - T_2 - 500 = 20\,000a$ (2)

Truck B:

$T_2 - 500 = 10\,000a$ (3)

Adding the three equations:

$8000 = 80\,000a$

$a = \frac{1}{10}$ m s^{-2}

From (3):

$T_2 = 10\,000 \times \frac{1}{10} + 500 = 1500$ N

From (2):

$T_1 = T_2 + 500 + 20\,000a = \mathbf{4000\ N}$

(ii)

Applying Newton's second law to the whole train:

$14\,000 - 2000 = 80\,000b$

$b = \frac{12}{80} = \frac{3}{20}$ m s^{-2}

Applying Newton's second law to truck B:

$T_2 + 4000 - 500 = 10\,000 \times \frac{3}{20}$

$T_2 = -2000$ N

The coupling between the trucks is under a **compression of 2000 N.**

(iii) $P - 4000 - 800 = 40\,000 \times b$

$P = 10\,800$ N

Total driving force is **10 800 N.**

How to solve these questions

Now you can apply Newton's second law to each component of the train.

Note that this equation is the same as applying Newton's second law to the whole system.

Sum of forces = total mass × acceleration.

You have not used equation (1), so you can now use it to check that your calculations are correct.

Remember that the acceleration will change with the increased forward force.

A negative tension means a compression.

Remember that there is an action/reaction force pair between the second engine and truck and B.

Apply Newton's second law to the second engine.

Answers

Q5

(a) (i) Applying Newton's second law to A:

$pmg - T = pm \times \frac{1}{2} g$ (1)

Applying Newton's second law to B:

$T - qmg = qm \times \frac{1}{2} g$ (2)

From (1): $T = \frac{1}{2} pmg$

(ii) Adding (1) and (2):

$(p - q)mg = \frac{1}{2}(p + q)mg$

$\mathbf{p = 3q}$

(b) (i)

(ii) Now solving $p = 3q$ and
$p + q = 12$

$\Rightarrow \mathbf{q = 3}$ **and** $\mathbf{p = 9}$.

How to solve these questions

Start with the force diagram and then apply Newton's second law to each object.

Essentially, the left-hand mass is three times the right-hand mass for an acceleration of $\frac{1}{2}$ g.

Since the acceleration is the same:

$p + q = 3 \times 4 = 12$

Answers

16 CONSERVATION OF MOMENTUM

Q1

(a) Momentum before collision
$= 3m + 0.5 \times 0$
$= 3m$

Momentum after collision
$= 1.4m + 0.5 \times 1.4$
$= 1.4m + 0.7$

So $3m = 1.4m + 0.7$

$\mathbf{m = 0.4375\ kg}$

(b) $I = 0.5 \times 1.4 - 0.5 \times 0$
$= \mathbf{0.7\ Ns}$

Q2

(a) Momentum before collision
$= 0.8 \times 200 \times 1000 + 0 \times 20$
$\times 1000$

$= 160\,000\ Ns$

Momentum after collision
$= 0.7 \times 200 \times 1000 + v \times 20$
$\times 1000$

$= 140\,000 + 20\,000v$

So $160\,000 = 140\,000 + 20\,000v$

$v = \mathbf{1\ m\,s^{-1}}$

(b) $I = 200\,000 \times 0.7 - 20\,000 \times 0.8$
$= \mathbf{-20\,000\ Ns}$

(c) $F \times 0.5 = -20\,000$
$\mathbf{F = -40\,000\ N}$

Q3 Total momentum before
collision $= 0$

Total momentum after
collision $= 48 \times 0.5 + 1.2v$
$= 24 + 1.2v$

$0 = 24 + 1.2v$

$v = \mathbf{-20\ m\,s^{-1}}$

How to solve these questions

First find the total momentum before and after the collision.

Then you can use conservation of momentum to form an equation to find m.

You can now use the formula $I = mv - mu$ with $m = 0.5$, $u = 0$ and $v = 1.4$.

You should calculate the total momentum before and after the collision and then apply conservation of momentum to find the speed.

Use the formula $I = mv - mu$ with $m = 200\,000$, $u = 0.8$ and $v = 0.7$.

You should know that the negative sign indicates that the impulse opposes the motion of the train.

Use the formula $I = Ft$, with $t = 0.5$.

As both are at rest the total momentum is zero before the collision.

The negative sign indicates that the skateboard moves in the opposite direction to the child.

123

Answers

Q4

(a) Momentum before collision
$$= 2x + 0.1 \times (-3)$$
$$= 2x - 0.3$$
Momentum after collision $= 0$
$$2x - 0.3 = 0$$
$x = 0.15$ kg

(b) Case 1: Both move in opposite directions.
Momentum after collision
$$= 0.1 \times 1 + x \times (-1)$$
$$= 0.1 - x$$
So $2x - 0.3 = 0.1 - x$
$$3x = 0.4$$
$x = 0.133$ kg (3 s.f.)

Case 2: Both move to the left at the same speed.
Momentum after collision
$$= x \times (-1) + 0.1 \times (-1)$$
$$= -x - 0.1$$
So $2x - 0.3 = -x - 0.1$
$$3x = 0.2$$
$x = 0.0667$ kg (3 s.f.)

Case 3: Both move to the right at the same speed.
Momentum after collision
$$= x + 0.1 \times 1$$
$$= x + 0.1$$
So $2x - 0.3 = x + 0.1$
$x = 0.4$ kg

Q5

(a) For the impact velocity U:
$$U^2 = 0^2 + 2 \times g \times h$$
$$U = \sqrt{2gh}$$
For the rebound velocity V:
$$0^2 = V^2 + 2 \times (-g) \times H$$

How to solve these questions

You need to describe the three ways that the particles could move.

Apply conservation of momentum.

Apply conservation of momentum.

Apply conservation of momentum.

You can calculate the impact velocities using the constant acceleration equations. You have to define one direction as positive, in this case this is the upward direction.

Answers

(b) $I = m\sqrt{2gH} - (-m\sqrt{2gh})$
$$= m\sqrt{2g}(\sqrt{H} + \sqrt{h})$$

Q6

(a) Momentum before collision
$$= 3 \times 3 + 1 \times (-1.5)$$
$$= 7.5$$
Momentum after collision
$$= 4v$$
So $7.5 = 4v$
$v = 7.5 \div 4 = 1.875\ \text{ms}^{-1}$
B changes direction.

(b) $I = 3000 \times 1.875 - 3000 \times 3$
$$= -3375\ \text{Ns}$$
Magnitude = **3375 Ns**

17 SAMPLING

Q1 Number sets 000 to 588.

Select three-digit random numbers.

Ignore repeats.

Ignore numbers greater than 588.

When 12 numbers have been obtained choose the corresponding sets.

Q2 Relevant prior knowledge is used to divide the population into strata. Random samples are taken from each of the strata usually in proportion to the size of the strata.

Stratified sampling should be used if relevant prior knowledge and sufficient resources are available.

How to solve these questions

You can use $I = mv - mu$.

You need to find the total momentum before and after the collision.

Then you can apply conservation of momentum. The velocity of B changes from -1.5 to $1.875\ \text{ms}^{-1}$ and so B has changed direction.

Now use $I = mv - mu$ with $u = 3$ and $v = 1.875$.

As you are asked for the magnitude, you should disregard the negative sign.

It is best to start with 000 rather than 001 but don't forget to finish at 588 (589 − 1).

You don't want to choose the same set more than once.

Be careful to exclude all numbers which are not on a set and to include all numbers which are on a set.

However if random sampling is not used the sample is called a quota rather than a stratified sample.

Answers

Q3 In a systematic sample items are selected at regular intervals, for example every 100th person entering a supermarket or every 50th item from a production line.

It is much less effort than random sampling. It is unsatisfactory if it leads to the items chosen being untypical of the population. For example if it is known that every 50th item on a production line is to be inspected special care may be taken with these items.

Q4 If a sample of employees of UK motorway service stations is required for a survey it would be impractical to select a random sample from the whole population. Instead a random sample of motorway service stations could be selected and then, say, ten employees selected at random from each of the chosen stations.

The advantage is that it is quicker as the interviewer only needs to visit a few service stations.

The disadvantage is that the employees of each service station are likely to be more homogeneous than the population of all service station employees.

Q5
(a) (i) **Yes,** each house would have a probability of $0.5 \times 0.5 = 0.25$ of being included.

How to solve these questions

It is often easier to give an example than to explain what you mean.

In most circumstances systematic samples are perfectly satisfactory.

A random element is retained but the employees to be interviewed are clustered in a few service stations.

In this context homogeneous means having similar views.

It is not easy to explain in words why the probabilities are equal. It is much easier to explain by making the calculation.

Answers

(ii) **No,** because two houses next door to each other cannot both be included in the sample.

(b) **Not a random sample of houses.**

Houses contain different number of residents on the electoral register so not all residents are equally likely to be included in the sample.

(c) Number residents from 00 to 62.

Select two-digit random numbers.

Ignore repeats and greater than 62.

When seven are found, choose the corresponding residents.

Q6
(a) (i) C
(ii) **A** random
B stratified

(b) **A** yes
B yes
C no

(c) (i) **A** gives all employees an equal chance of being included in the sample.
(ii) **C** is a much quicker and easier method of obtaining a sample.

(d) Employees of different factories may have different views on working conditions. **B** ensures that employees of all factories are fairly represented in the sample.

How to solve these questions

Again it is easiest to explain with an example.

You could also make points about some residents of a house being more likely to be at home than others.

An alternative acceptable answer for C is 'it depends on what convenient method is chosen.'

18 PROBABILITY

Q1
(a) 0.23 — $1 - 0.64 - 0.13 = 0.23$
(b) (i) 0.0529 — 0.23^2
   (ii) 0.294 — There are two ways of this occurring: $2 \times 0.23 \times 0.64$
(c) (i) 0.262 — 0.64^3
   (ii) 0.295 — The probability of not paying by cheque is $1 - 0.13 = 0.87$
   $3 \times 0.13 \times 0.87^2$
   (iii) 0.115 — $6 \times 0.64 \times 0.13 \times 0.23$

Q2
(a) 0.343 — $0.7 \times 0.7 \times 0.3 + 0.7 \times 0.3 \times 0.6 + 0.3 \times 0.6 \times 0.7$
(b) 0.399 — $\frac{1}{6} \times \frac{2}{5} + \frac{5}{6} \times \frac{4}{5}$

Q3
(a) 0.733 — P(windy and target) = P(target) × P(windy|target)
(b) 0.0909 — $\frac{1}{6} \times \frac{2}{5} = 0.7333 \times$ P(windy|target)

Q4
(a) 0.008 — 0.2^3
(b) 0.142 — $0.2^3 + 0.35^3 + 0.45^3$
(c) 0.0563 — P(all water and all same) = P(all same) × P(all waterfall same)
P(all water and all same) = P(all water)
$0.008 = 0.142 \times$ P(all waterfall same)

Q5
(a) 0.667 — 80 out of 120 students passed.
(b) 0.392 — 47 students were < 20 and passed.
(c) 0.967 — All but 3 + 1 = 4 students were either < 20 or didn't fail (or both).
(d) 0.733 — There were 45 students aged 20 or over (Q'), of whom 33 passed.
(e) 0.25 — $P(F) = \frac{45}{120}$ $P(F \cap R) = P(F) \times P(R)$, F and R are independent.
(f) 0.3 — $P(F \cap Q) = P(F) \times P(F \mid Q)$
(g) 0.133 — $P(S \cap F) = P(F) \times P(S \mid F)$

Q6
(a) (i) 0.01475 — $0.01 \times 0.98 + 0.99 \times 0.005$
   (ii) 0.664 — $P(D \cap +ve) = 0.01 \times 0.98 = P(+ve) \times P(D \mid +ve)$
(b) 0.653 — $0.6644 \times 0.98 + (1 - 0.6644) \times 0.005$

19 BINOMIAL DISTRIBUTION

Q1
(a) 0.733 — B(20,0.09) You can read the answer from tables.
P(4 or fewer) − P(3 or fewer)
(b) 0.0703 — B(8,0.43) tables won't include this, so you will have to use the formula.
P(0) + P(1)
1 − P(7) − P(8)

Q2
(a) 0.253 — B(10,0.3), which you can read from tables.
P(4 or fewer) − P(3 or fewer)
(b) 0.0784 — P(5 or fewer) − P(1 or fewer)
(c) 0.986 — B(10,0.7) but this is not in tables.
4 or fewer 'failures' → 6 or more 'successes' → 1 − 5 or fewer 'successes'.

Q3
(a) 0.850
(b) 0.200
(c) 0.803
(d) 0.0473
6, 2.05 — $np, \sqrt{np(1-p)}$

Q4
(a) (i) 0.210 — B(16,0.15): 1 − P(3 or fewer), but not all binomial tables will include this.
   (ii) 0.131 — P(4 or fewer) − P(3 or fewer)
(b) (i) 0.934 — B(5,0.2101): you will have to calculate this. Use as many significant figures as your calculator will give you for p.
   (ii) 1.05 — 5×0.2101 (The expected number is the mean.)

Q5
(a) 1.525, 1.48 — From the calculator; the standard deviation with divisor n giving 1.47 is also acceptable.
(b) 0.381 — 61 out of 160 jumps were disallowed.
(c) 1.525, 0.944 — $np, np(1 - p)$
(d) No, the standard deviation predicted by binomial $\sqrt{0.944}$ = 0.971 is a long way from the observed value of 1.48.
(e) Some competitors are more likely to have jumps disallowed than others → **p is not constant.**

Q6
(a) (i) 0.0905 — B(25,0.3) which you can read from tables.
   (ii) 0.147 — 19 not agree → 6 agree, P(6 or fewer) − P(5 or fewer)
(b) (i) **Yes, n fixed, p constant**
   (ii) **No, n not fixed**

20 NORMAL DISTRIBUTION

Q1
$z = -1.6$
P(X < 32) = 0.0548 (3 s.f.)

P(X < 32) = 1 − 0.945 20 = 0.0548

Q2
$z_1 = -1.5$, $z_2 = 2.0$
P(component meets the specification) = **0.910**

P(component meets the specification)
= 0.977 25 − (1 − 0.933 19) = 0.910

Q3
(a) $z = -0.5454$
The proportion with blood nicotine levels below 250 is
1 − 0.707 = 0.293

An answer based on rounding z to 0.55 would have been accepted.
Interpolating between $z = -0.54$ and $z = -0.55$

(b) The blood nicotine level exceeded by 20% of smokers is 310 + 0.8416 × 110 = **403**

Q4
(a) $z = 1.5$
probability < 5800 = **0.933**

(b) $z_1 = -1.75$, $Z_2 = 1.5$
P(lifetime between 4500 and 5800 hours) = **0.893**

P(lifetime between 4500 and 5800 hours)
= 0.933 19 − (1 − 0.95994) = 0.893

(c) For the 84th percentile,
$z = 0.9945$
The lifetime is **5600 hours** (3 s.f.).

Remember that the 84th percentile is the value which exceeds 84% of the population.
The lifetime = 5200 + 0.9945 × 400 = 5597.8 hours.
Round to three significant figures.

Q5
(a) (i) $z = 1.25$
P(time spent in the library will be < 90)
= **0.894**

(ii) $z_1 = 0.25$, $z_2 = 1.25$
P(spending between 60 and 90 minutes) = **0.493**

P(spending between 60 and 90 minutes)
= 0.894 35 − (1 − 0.59871) = 0.493

(b) The maximum time a user can spend in the library is 60 minutes. A model with mean 65 minutes could not apply.

(c) 99.9% of normal distribution exceeds $\mu + 3\sigma$. The model would be plausible if a user has at least 65 + 3 × 20 = 125 minutes in the library. **The user must enter by 6.55 p.m.**

A less cautious answer based on 95% or even 90% of the distribution would be acceptable.

Q6
(a) (i) $z = 1.4$
The proportion of large eggs is 1 − 0.919 24
= **0.0808**

(ii) $z_1 = -0.6$, $z_2 = 1.4$
The proportion of medium eggs is **0.645.**

The proportion of medium eggs is 0.919 24 − (1 − 0.72575)
= 0.645.

(b) $z = 2.0537$
The minimum chest measurement for extra large is **111.3 cm.**

The minimum chest measurement for extra large is
101 + 2.0537 × 5 = 111.3 cm.

(c) The proportion of people needing small T-shirts is
1 − 0.72575 = 0.27425, the median of these must exceed 0.27425/2 = 0.137125 of the population.
Median measurement of small shirts is 101 − 1.09 × 5
= **95.5 cm.**

You should find this answer, using interpolation $z = -1.09$.
Rounding the probability to 0.14 and obtaining $z = -1.08$ is also acceptable.

21 CORRELATION AND REGRESSION

Q1 **0.868**
Heavier ewes tend to have heavier lambs.

You can use your calculator to find this answer.
Positive value close to 1 shows strong association.

Q2
(a)

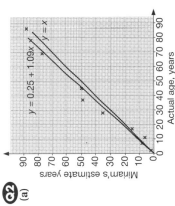

(b) **y = 1.09x + 0.246**

From calculator

(c) **Miriam's** estimates lie close to the regression line. This line is a little above the ideal line $y = x$. **She tends to overestimate slightly.**

You can see this from your diagram.

Answers

Q3

(a)

For countries with low income per head a small increase in income is associated with a large reduction in infant mortality. This effect is less marked for countries with a higher income per head.

(b) (i) **−0.739**

(ii) The relationship appears not to be linear so the product moment correlation coefficient is **not appropriate**.

(c) **−0.936**

(d) Both show an association between low income and high infant mortality. Spearman's rank correlation coefficient is much closer to 1 because although the relationship is not linear, the ranking of the countries on income is almost the reverse of the ranking on infant mortality.

How to solve these questions

The non-linear relationship which is clear from the scatter diagram is the key to this question.

From calculator

First rank both sets of data. There are no ties so your calculator will give you the same answer as using the formula based on Σd^2.

Answers

Q4

(a)
$$b = \frac{484 - \dfrac{143 \times 391}{15}}{2413 - \dfrac{143^2}{15}} = -3.089\,86$$

$$a = \frac{391}{15} + \frac{3.089\,86 \times 143}{15}$$
$$= 55.5234$$

$$y = 55.5234 - 3.089\,86x$$

$$h - 100 = 55.5234 - 3.089\,86(s - 20)$$

$$h = \mathbf{217 - 3.09s}$$

(b) The slope of the line is an estimate of the **reduction in life** (hours) for each revolution per minute increase in speed.

(c) **125 hours**

Q5

(i) *See diagram opposite.*

(ii) $y = \mathbf{7.36 - 0.000\,222x}$

(iii) The intercept estimates the diesel consumption of a lorry with no load.
The slope estimates change in diesel consumption for each additional kg carried.

(iv) 30 000 kg is outside the range of the observed data.

(v) Danny's diesel consumption is always less than predicted by the regression line.

(vi) The load may affect diesel consumption but diesel consumption cannot affect the load.

How to solve these questions

Unfortunately the raw data has not been given so you will have to substitute in the formulae instead of using your calculator directly.

Keep as many figures as possible for these calculations.

Now substitute for x and y to obtain the equation in terms of h and s as required.

This is the required equation so now round your values to three significant figures.

Substitute $s = 30$ in the equation.

Since diesel consumption is measured in km/litre this is undesirable.